MEGA SEX

Dr. Francine Beck

Dunhill Publishing Co.
New York

Mega Sex

A DUNHILL TRADE PAPERBACK
Published by: Dunhill Publishing Company
A division of the Zinn Publishing Group
ZINN COMMUNICATIONS/NEW YORK

ISBN: 0-935016-08-2

Printed in the United States of America

Library of Congress Cataloging-in-Publication Data

Beck, Francine.
    Mega Sex / Francine Beck.
      p. cm.
    Includes bibliographical references.
    ISBN 0-935016-08-2 (pbk. : alk. paper)
    1. Impotence—Prevention. 2. Impotence—Alternative Treatment.
I. Title.
RC889.B425 1997
616.6'92—DC20
                                      96-26763
                                         CIP

I dedicate this book to my Maker who so generously gave me the talent, perspicacity, persistence and love to reach out to my readers . . . and to John Pisano, my best friend, whose insights continually amaze me!

This book is not intended as a substitute for medical advice. The reader should regularly consult a physician in matters relating to his or her health and particularly with respect to any symptoms that may require diagnosis or medical attention.

This book contains many references to actual cases the author has encountered over the years. However, names and other identifying characteristics have been changed to protect the privacy of those involved.

TABLE OF CONTENTS

FOREWORD

Dr. Beck has written a breakthrough book in her field. Rather than present a Universal Formula for Sexual Dysfunctions, her own personal views, a regurgitation of other books, or repetitious case histories, she presents the reader and his/her partner with individual help by implementing the latest New Age Techniques accumulated after years of research, not only in libraries and bookstores, but in the laboratory of the world.

What MEGA SEX offers the reader is a window on coping mechanisms for Sexual Dysfunctions. You and your partner can work on your own personal emotional blockages. How often does repressed anger or fear cause anxiety and sexual dysfunction! How can a person cope with these emotions and overcome them. Coping techniques are described both verbally and visually to deal with these blocks. Here is a book that can be used individually or by both partners in a relationship. We see how to present a sexual scenario to a partner with conservative ideas. Imagery and other exercises help us to perform our Personal Best in the Sexual Field.

Partners at different stages create stress in a relationship. Relationships come in all shapes and sizes and can be worked by one or both part-time, full-time, or once a month. Dr. Beck shows how Masters and Johnson Techniques, BioEnergetics, Imagery, Polarity, Cranial Cymatics can be used to release blockages. Methods and exercises are described in detail and illustrated with diagrams to reach both right and left brain functions. Here is a book both men and women can use to improve their own or their partner's sexual performance.

In MEGA SEX, Dr. Beck has drawn from her own unique experiences after working for years with many kinds of dysfunctions and sexual problems. No other book does what this book has done. People from all over the world have come to her for help. This is a universal problem found in all societies. It speaks to all nations, races and religions. This book is the best attack on Sexual Dysfunction I have seen since Masters and Johnson.

The Editor

CHAPTER ONE

WHY CAN'T WE CRY?

CRYING HELPS MY SEX LIFE? Yes! The physical and emotional expression when sobbing releases endorphins from the brain which improve your sex life. In my over seventeen years as a therapist, I have had every possible profession come in to me—doctors and lawyers and priests and rabbis and construction workers and judges—you name it, every kind of possible profession. But, I've never (in all these years) had an actor or a woman or a Latin as a patient. Why? What is the correlation? First of all actors provoke and re-enact their feelings by concentrating on an angry or sad or joyful moment, pinpointing it and experiencing it physically and emotionally, and then letting go. It may seem artificial or mechanical, but it works to help the individual experience / confront the emotion and then let it go both physically and mentally. By physically provoking anger and grief, hate, love, etc.—any of these emotions—the actor learns to see it clearly, act it out and start to get rid of it. People who suppress emotions day after month after year will discover that the mind hides it and the body begins to absorb it. It doesn't go away. It becomes cancer, arthritis, heart disease, psychosis and sexual dysfunction.

When we consider that 95% of the mind is subconscious and only 5% is conscious, what do you suppose happens to this anger and grief, etc.? When the unconscious needs to store it, it causes everything from colds to cancer; neurosis to psychosis. When an injustice is done us, the 5% conscious minds says, "Oh, it's okay, I'll get over it," but the 95% unconscious mind says, "I'm hurt, deal with me." Imagine how much of it is stored away in the mind year after year after year. If we do not help to get rid of it every day, it will cause impotence and premature ejaculation.

Women, Orientals, Latins

Women seldom have sexual dysfunction. They are permitted in our society to express anger and grief. It's not okay for men to cry. Notice at a funeral, men are conditioned to be tough, macho and never shed a tear. This is so sad, for they harbor all their pain and suffer with premature ejaculation and impotence. Did you ever see an unenthusiastic Latin? Travel to Spain, Italy, or South America. Whenever a situation arises, be it positive or negative, the people usually express it, feel it, and let it go. If you watch the TV-cable shows, all the Latinos are very joyful or in tears. They are not conservative in their display of emotion. Observe: there's one TV channel that has all Oriental people on it. It's interesting, that if you observe the Orientals, their culture teaches them not to show their feelings . . . to suppress them. 90% of my patients are Oriental and I only treat sexual dysfunction. That's telling you something, right? I'm not encouraging everyone to begin punching your enemy or bursting into tears at every moment you feel this need. In Chapter VII, we'll learn how to give 15 minutes a day to expressing feeling and letting go through Bioenergetics.

Why Can't Men Cry?

Most men won't acknowledge rotten feelings let alone confess them. People don't understand that emotional responses are involuntary. They think we're somehow responsible when we get angry or sad—as if we were willed to feel that way. Of course, one way of dealing is going into denial or rationalizing. But, because emotions are our personal feedback system, learning how to recognize them in everyday life will leave us better prepared to cope when trauma strikes. Constantly repressing emotions may wreak havoc on the immune system, and if the immune system is impaired, it may not be able to guard against disease. Endorphins (hormones with healing ability) are secreted by the brain when emotions are expressed. Just how passionate can you get over a sad movie, song, or poem? Do you feel choked up inside, but never or seldom ever express it in big tears or explosive laughter or uncontrollable rage? Can you truly feel the pain of a blind person stumbling or a mother losing her child or a child losing her dog? Can you let your grief feel complete relief through big tears? Can you let your heart break? Or, do you stop yourself and say, "Oh, I'll get over it." Your unconscious mind will harbor this negative feeling, so why not let it go when you feel it? Surely as a male child, our parents, teachers and other authority figures scolded us when we wanted to scream in anger or cry when our hearts were broken. "Don't scream, don't cry, control yourself, be strong, be a man, be a man." How sad. Why are we taught to suppress when it is natural to release or express? Do we want approval? Is this why we suppress anger and grief? If mother passes away and the average male wants the approval of his wife, children, or society, can he cry? Will he cry? Society dictates that men are not supposed to cry, so he shows no emotions. Not if he wants the approval. That's what he's looking for, so he controls himself and his negative feelings. They are stored away in a big bank of emotional and physical disease. Tears can represent grief, joy, anger and frustration.

Why Cry?

Evolution has shown us that virtually everything our bodies do happens for a reason. Humans are the only animals that cry. Man cries for emotional and physical reasons. It's one of the few things that sets us apart from the other creatures. There must be some survival value in crying. Perhaps it is a stress releasing mechanism. It is an excretory process that aids in the removal of substances that cause stress. It has been proven that emotional tears differ from tears shed while peeling onions or inhaling fumes. Lacrimal tears have the ability to remove manganese. Manganese is a mineral directly related to stress found in tears. There are also several hormones found in tears. One of them is adrenocorticotropic (ACTH). It has been shown in blood levels to be the most sensitive indicator of stress. Animals also have ACTH and when they are subject to stress, the levels of ACTH can get high enough to cause brain damage in terms of memory, and since they don't cry, they can't release it. When human non-criers such as men cannot shed tears, chemical byproducts become a health hazard. Repressing emotional tears is a source of great tension and diversion of psychic energy. This distorted reaction has been known to cause such illnesses as arthritis, colitis, headaches, insomnia, and a whole range of psychological problems.

Delayed expression of grief can be expressed even years later through Bioenergetics which I discuss in Chapter VII. This treatment is very effective when tears have been suppressed due to trauma, physical illness, or machismo. Diary keeping appears to help the griever since it constitutes a form of cathartic expression. It appears that attempts to control feelings affect actual feelings. Grief tends to be stronger or more complex in those who use more words. Either the amount one writes or the richness with which one expresses feelings may influence the intensity or complexity of feelings. Vocabulary may enhance communication more than it does feelings. This applies to authors, playwrights, poets, and composers. Why have we learned to push back the pain behind a smooth exterior? Is it a sign of weakness to express one's hurt? We are conditioned to be brave and smile no matter what. By staying busy we appear happy. We do this through avoidance, prayer, activity, nervous laughter, and inactivity. Showing that your heart bleeds and letting the tears flow is not weak. We must unlearn this control of feelings. A society which teaches us to always be on guard, produces sick people.

Case History:

I know of a couple married for forty years who thought they knew one another. The husband had always communicated his joy but very little of his fears. He had to be strong and in control. He had suffered bursitis in his shoulder for many years, and even wore his arm in a sling when the pain was severe. His wife was accustomed to hearing him occasionally mention his pain. On the night he died he was in severe pain, but he was too proud to tell her how bad it was. He was having a heart attack and would not even let her call the doctor or ambulance. He thought it was just bursitis; was too proud and did not want to admit that he was in severe pain. When he continued to mention the pain, she insisted on calling the doctor even though it was 3 o'clock in the morning. He kept holding down the phone receiver. She escaped to the kitchen extension and called the doctor; the doctor said to call an ambulance right away. She was terrified that she had not realized the seriousness of the situation. Because the husband had been so conditioned to bear the pain *like a man*, he died in the hospital of a heart attack.

Suppressing Grief

Many people work so hard at controlling grief when a good cry would relieve the pain. After this catharsis they could go on with their work, play, learning—living. The literature on grief commonly treats social isolation as a symptom of poor grief work, but the social isolation may be the cause of poor, or at least relatively slow grief work. The fact that people who live alone tend to be more depressed may be pertinent to their lack of social interactions to distract them from their grief. If social interactions can reduce the pain of grief, why would people choose to be socially isolated? Many people cannot show their grief because they feel it is too painful. Every scene and every face serves only to remind them and embitter every part of their lives, making socializing impossible. Why then not grasp these opportunities to cry?

Avoiding the expression of grief is a self-defeating activity; the sufferer never really deals with the pain. With this avoidance, one becomes vulnerable to everything from colds to cancer, from neurosis to psychosis. Because we are conditioned by parents and society not to express our feelings, the mind and body rebel since their innate reaction is to cry. Joining a bereavement group is often beneficial. The mutual expression of grief may lessen the pain when one realizes many others are also suffering a loss. At these meetings loss is often meaningful through sharing. Many people who write feel grief with more intensity or complexity. The *expression* of grief seems to lead to a snowballing of grief. The more one expresses, the more one feels and experiences a catharsis. This catharsis is achieved through Bioenergetics (Chapter VII), where there is an association between verbal fluency and grief.

When anger is expressed at the same time an injustice is done, the physiological phenomena causes many illnesses. When expressed through Bioenergetics as covered in Chapter VII, the physical stimulation is not as intense and is healthier for the body. At the time of an anger inducing experience, the person reacts immediately. There are numerous physical results. The volume of saliva and the quantity of hydrochloric acid secreted by the stomach is increased. This increase constitutes a cholinergic stimulation and the gastric mucosa becomes hyperemia and engaged. Stimulation of the hypothalamus causes signs of angry behavior which also prompts muscular vasodilation mediated by automatic nervous innovation, which can also be observed when nausea accompanies the anger response. It appears that in the anger response both parasympathetic and sympathetic discharges are operative in the stomach. Under the influence of anger, the heart rate increases rapidly. The tachycardia is mediated through liberation of adrenergic hormones. Anger responses produce blood pressure elevation, and frequent recurrences of such emotional reactions as hypertensive episodes, may predispose the subject toward the development of essential hypertension. The anger response is followed by an absolute decrease in lymphocytes in the circulating blood. The elevation of free fatty acids in the blood were reported during the anger response.

Delaying the Reaction

During the anger response, much emphasis has been placed on electroencephalographic abnormalities. These psychological reactions were not present when anger was provoked in Bioenergetics. The desired effect from Bioenergetics was a physically and emotionally healthier person. Allowing the vulnerability to evolve rather than acting upon it immediately appears to be a much more rational expression of anger. Though the more natural expression is immediate, the delayed expression is more healing. When the subject is not traumatized with both the anger response and grief reaction, it becomes easier to release the emotion and prevent both mental and physical illnesses. When there is acute anger and grief, and the subject is permitted to acclimate to the shock, he/she can act out the rage, depression, panic, and pain, hostility, aimlessness, restlessness, irritability, etc. A delayed rehearsed expression of a painful experience is necessary for emancipation. So we reawaken all the painful occurrences, focus in on them, reenact them, and have a magnificent catharsis. A divorce, a death, a strong relationship, a disappointment, a termination of a relationship or loss of a job or socialization, change in the body (losing or gaining quite a bit of weight), or severe illness, operation, accident and a host of shocks to the system will awaken strong hate and love, thus causing excessive anger and grief. At the time of their occurrence, the shock to the subject may leave him unable to react since the traumatic event usually paralyzes the person. Interestingly, there is a matter of nature saying, "this is too much for the body and mind to handle." We, the body and mind, will lay low until we have absorbed this, can deal with it and survive unharmed such as perhaps reacting in paralysis or amnesia. The lower animals cannot reason this way and react immediately. Of course they must do so truly, or perish. But man can delay his reaction, thereby preserving his mind and body, and possibly warding off illnesses. The body and mind are not prepared for severe shock and begin to fight the rape to its system by depressing its immunity as a defense or perhaps because of its need to acclimate. This is where Bioenergetics come in.

Dr. Alexander Lowen, originator and author of hundreds of books on Bioenergetics, explains how this occurs. He has cured many patients of terminal illnesses through Bioenergetics as well as patients of psychosomatic illnesses such as colitis, asthma, arthritis, etc. Physical and mental illnesses appear when the body's immune system is not prepared to handle the shock. Sometimes paralysis and depression are a way the body gives itself a chance to delay the shock and deal with the problem when it feels stronger. Feelings of hopelessness, helplessness, and amnesia are common ways the body deals with shock in order to give it adequate time to react and perhaps prevent severe illness. The usual first stage of anger is shock, frantic activity, severe anxiety. The second stage is usually manifested by withdrawal, regression, disbelief. Aggression is often followed by physical activity which can put the aggressor in jail, pain, etc. When anger is not expressed, then acted out daily, a few months later the danger is often less severe and handled intelligently with no repercussions. Through Bioenergetics, it is done privately (see Chapter VII). No one is hurt, and the one offended is relieved of his anger with no negative actions to either. This delayed reaction is the better attempt to restore equilibrium in response to the change in external milieu. This is a mechanism of conservation of energy until a new resource or supply is available. Imagine all the crime which could be averted if people were taught to delay their anger—thereby softening the blow and tempering it with healthy physical activity through Bioenergetics.

While further verbalizing the pain, the angered person can, with the addition of physical percussive expression, resolve the rage. Sexual dysfunction and angry or sad experiences are so emotional in the human experience, that the specific effect of rage and grief on sexuality becomes obscured. The concept of self as a sexual person is as vulnerable to grief and anger as our sexual roles; responses to stimuli and relationships. Illnesses which arise from shock are often the cause of sexual dysfunction. Grief in response to physical illness and disability, is magnified as the individual must simultaneously face threatened sexual performance and altered sex coded role behavior. The immediate adverse effect of some illnesses or some disabilities caused by shock on physical sexual expression cannot be underestimated. To avoid these illnesses, the person must delay his rage and sorrow until he can react through Bioenergetics. The many illnesses caused by the prompt response to shock affect the endurance and erectability of the male. The effect of grief on rage to the physical sexual response will also reduce the desire. The depressive aspect of grief on rage has the potential of lowering libido and leads to feeling unworthy of sexual pleasure. Gradual controlled expression of severe emotions by the experienced practitioner of Bioenergetics will eliminate illness and allow the subject to perform with harmony and satisfaction. The sexual desire essential to the physical expression is terminated or greatly reduced when a shock is experienced by the system. Such sexual frustration is felt on both the emotional and physical level. When these illnesses occur after shock, the male's ability to provide for his family is closely associated with his ability to perform sexually. Where grief or rage are demonstrated by rigidity and restrictive patterns of behavior, adaptation is severely affected by sudden emotional expression and extreme sexual problems.

CHAPTER

TWO

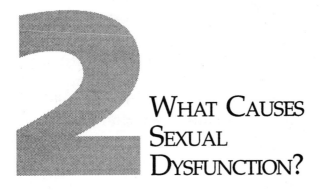

WHAT CAUSES SEXUAL DYSFUNCTION?

Do you think it is your age or inexperience or economic situation? They're not responsible for your sexual dysfunction. What is?

Impotence and Premature Ejaculation

Impotence refers to the inability to achieve or sustain an erection. There are approximately twenty-two million people in the United States alone who *reported* their impotence. So, it's quite common. Many patients begin treatment with a negative attitude engendered by doctors, friends, society, etc. One of these misconceptions is that they are sixty years old and it's time to let up on their sex life. After all, didn't Uncle Joe say when you get older you don't want it anymore. And didn't Dr. Smith say that your sex parts, like the rest of you, slow down when you get (as they say) over the hill. And whoever coined that expression "over the hill?" To be over anything good is so negative to the mind. The word "retired" is so final. When we decide to slow down, not work, not contribute to society, the mind performs a somersault. Suddenly after working steadily, this fruitful, productive organ is told to stop. Such waste; such negativity. And you wonder why the sex parts also slow down. Instead of calling it retirement, let's rename it, let's call it resparking, reliving, revitalizing, regenerating, renewing, or resurrecting. Resurrection instead of retirement. This positive word would keep us from feeling over the hill, finished, ending, slowing down, out of it. If your mind thinks it is finished, go try and tell your sex organ that it's not so. After all, the penis is an appendage of the brain.

Quite a number of impotent men have a great fear of being overly assertive, aggressive, penetrative. They have learned that such qualities are undesirable. They often have guilt regarding their rage about women. Many fear that their penis is a loaded bomb which will explode inside a woman causing massive damage.

Most impotent men haven't let themselves experience their true feelings of helplessness, powerlessness, or rage and hatred towards women. They have denied their feelings. These feelings have taken a bodily dimension in the form of sexual impotence.

Impotent men often have fantasies that their penis is a dangerous weapon which must be kept hidden. Conflicting messages bombard men: Be a man. "Don't be a wimp." "Be on guard lest you betray your masculinity with a feminine gesture, remark, mannerism, or emotional outburst." But on the other hand: "Don't be brutal. Don't be savage. Don't be aggressive. Be reasonable." These opposing injunctions pressure many men, squeezing them harder and harder until they feel desiccated, lifeless and impotent. So men try and do a tightrope in the middle, but throughout there's an anxious voice inside: "Have I got it right?" They can easily end up feeling castrated, rendered impotent by a whole barrage of demands and injunctions. We have also seen the macho man as a castrated man. His very caricature of maleness renders him powerless in many circumstances. He is terrified of his feelings. He may feel afraid or panicky or he might want to collapse. But all of these possibilities are ruled out as he keeps up his fierce disguise of unbending strength. Thus, in this case, castration means being deprived of their more feminine feelings. The man who cannot let a woman be on top in bed, who has to retain control sexually, who can't weep, who vehemently rejects any homosexual fantasy, is in fact castrated quite as much as the man who can't get angry. External misogyny connects with the denial of the man's inner femininity. In hating women, such men hate themselves. He cannot let himself shake with fear or weep with exhaustion or grief. It's hard for him to cry. This is the true castration of modern men. They function from the neck up. Their bodies are deadened and therefore their feelings are unavailable to them. It's hard for them to cry, to be warm, to melt, to love. A truly powerful man is sensitive, emotional, is able to cry, can sometimes admit he can't cope, can allow himself to be passive at times and can accept feeling powerless, but can also accept feelings of rage, brutality, sadism. Indeed, the strong and healthy male embodies the whole spectrum of emotions which human beings are privileged to enjoy.

Case History:

I'd like to tell you about a patient I had, a young man. Tony was a forty-nine year old, handsome, sensitive, well-employed gentleman who was sent to me for impotence. He still lived with his parents. A colleague referred him. She had been working with him for six years in therapy and had performed two exorcisms on him. Let's talk about the exorcisms. They cost $7,000 each. I witnessed one. It was very elaborate, with a head man who I assumed to be a priest. He gave commands to about four people. He was dressed in black leather and came into the room rapping. These people went into gyrations when spirits entered their bodies. I believe they were actors. It was very ritualistic and I felt it was ridiculous and phony. In the end it didn't help him anyway. The exorcist also sold him a vase with spirits in it costing another $6,000. Now this poor fellow came to me in desperation. He had spent all his money, but was still impotent. What happened was that he was so afraid of the spirits in the vase, that he moved out of his house. He was living in the streets. That's how desperate he was. Tony was a special case; a virgin, and painfully shy. He had never masturbated. He was afraid that the spirits would see him and watch him and he was ashamed. I told him he had to masturbate. He needed a release. I said, "When you go home, close the door and you tell the spirits to wait in the other room." He never thought of that and neither had the exorcist. When he went home he tried this. He locked the door and he told the spirits to wait and he masturbated. This little advice gave him the courage that he needed. He returned to me so thrilled. The next step was to get him to actually engage in sex with a real live girl. I suggested he attend several singles parties. He finally mustered up the courage to speak with the ladies. I instructed him regarding what to say and he eventually dated one and is now happily married.

I have so many very successful businessmen, celebrities, politicians, doctors, lawyers, musicians, etc., who actually believe they can't do anything about impotence. It sounds pretty extreme to go to an exorcist for impotence and it certainly is. But, it's better than medication or operations. Some doctors would have you coming for treatment to keep them in their Mercedes, or for an indefinite period of time, to keep them wealthy. They have you coming back week after week, after month, after year, yet they still cannot offer a cure. One patient confessed he had gone to a psychiatrist for nineteen years. Still he was impotent. How preposterous. If a doctor can't help you after a few months, find another doctor!

Premature Ejaculation

According to the latest statistic, seventeen (17) million people in the United States alone suffer from premature ejaculation. The clinical definition of premature ejaculation is the *"inability to control your ejaculation for fifteen minutes after you enter the vagina,"* or other orifice. Another definition is *"not being able to control it until your partner orgasms at least 60% of the time."* When I say average, please know that I mean about 65%. The average premature ejaculation patient is nervous, anxious, impatient, and confused. He is a nail biter, procrastinator, perfectionist, skeptic—totally unaware of his feelings and how he feels about himself. He dislikes being rushed and is angered and annoyed if he has to wait ten minutes to see me. There's always something he has to rush to. He feels compelled to be in perpetual motion. He is so restless. It's difficult to get through the first exercise of my program which is only two minutes long. Fortunately, after this exercise, he is much calmer and I was able to proceed with the rest of the treatment. In Chapter VI on treatment, I will elaborate on what I feel the causes of premature ejaculation and impotence are. After sixteen years of research, observation, and study in the field of sexual dysfunction, I've discovered that the chief cause of both premature ejaculation and impotence is tension, which is also covered in detail in Chapter VI.

Certainly the tense person will always experience failure in his sex life. Isn't the penis an appendage of the brain? It stands to reason that if the mind says "I'm full of stress," the body says "I can't function well without your cooperation." Thus, the prime and most important need is to engage in techniques such as imagery, which causes you to relax the mind and the body in order to perform at your best. For premature ejaculation, we must distract the mind while having intercourse by applying a series of fun exercises, which along with several others (explained in Chapter VII), eventual resolution of the problem is assured. These exercises will allow the patient to gain control of his orgasms, so that he will not ejaculate until he wishes. It is very important for the premature ejaculating patient to masturbate at least daily. Why? When one urinates in the morning, the urine comes out with great force since the bladder is so full, and there's pressure supporting this function. So it is with premature ejaculation. In the male genitalia, when there is pressure on the amount of semen in the seminal vesicles, the semen gushes out much more rapidly. The sperm needs a daily outlet in masturbation. This is also excellent and necessary for the man with impotence. After all, what happens to a muscle if you don't use it? The muscles in the penis need activity or they will become flaccid and atrophy. Watch the people in a gym. Why are they so muscular? Because they use their muscles. Use it or lose it! Daily masturbation is also beneficial in its rewards to all men, since the frequent use of this exercise guarantees a firm penis in later years. I've treated men in their eighties with much success. Once we get past the misconception that age is why they are totally or partially impotent, we can proceed with great success. Unfortunately many men have been told by doctors, friends and spouses, that it's time to slow down or stop altogether. Once they become aware that this has often caused their impotence, and that they can go on having sex until the last day of their lives, they begin to believe it and can then function in any capacity that they wish.

The physical side of premature ejaculation must also be treated. In most instances, 70% of the cause of sexual dysfunction is psychological and 30% is physical. Temporary impotence can occur when the patient is taking antihistamines, depressants or tranquilizers. Longer term impotence can occur when the patient is taking medication for hypertension, heart disease or ulcers. It all depends upon the medication and dosage. Sometimes, if their medical doctor feels it is safe to decrease the dosage or change the medication, it will bring back a stronger erection. In the case of heart disease or hypertension, I will suggest that the patient request a complete physical checkup and ask their medical doctor whether his medication can cause this impotence. I often work with the doctor to equalize the medical and psychological treatments. For heart disease, my experience has been that the beta blocker shows a better outcome, as it seldom has a side effect. There can also be a hormone imbalance, too little testosterone, too much prolactin. If this is so, I will suggest that the patient first be treated by his MD, and return to me when his balance is normal. Even though the patient and the MD may feel that a disc problem will interfere with sexual functioning, I have never found this to be true. Patients with a history of bladder, prostate, or testicle cancers are asked to check with their MD to find out where they feel there is a possibility that any of these may cause impotence. I have been successful in treating as many as 80% of these patients. In my experience, the hernia or its repair, has never caused a permanent sexual dysfunction. The hydrocele or the varicocele, or their corrections may cause great discomfort, but not permanent damage to sexual functioning. In the case of spermatocele or cyst of the epididymis, no notable change will be detected in ejaculation. Surgery is usually not advised and it has no effect on sexual function. Mumps or torsion, or its treatment, will not render the patient impotent or premature. Cancer and other serious maladies will affect sexual performance depending on the severity of the disease. Such illnesses as Ureaplasma, Trichomonas, Herpes, Syphilis, Gonorrhea and Lymphogranuloma Venereum will all respond to drugs and will not affect sexual prowess. Statistics and information about AIDS and Hepatitis are continually changing and it is difficult to have a balanced perspective in the midst of an epidemic.

Impotence can occur when a patient has suffered an accident, resulting in a pelvic fracture which could leave him with injuries to the nerves, blood vessels and so on. This prevents the blood from flowing to the penal area. This is the only area where I have had difficulty in treatment, although in 30% of the cases, Polarity (see Chapter VII) has been effective in unblocking negative energy, thus bringing an adequate blood supply to the penal area.

Premature ejaculation has always been treatable, and as long as the patient does his homework religiously, there is no exception. I've had 100% success in treating premature ejaculation. The different forms of treating impotence are surgery, prosthesis, surrogate, the list goes on. Of all these approaches, surrogate therapy seems to be the least successful. A patient with a sexual dysfunction is by nature already uneasy with the opposite sex. This is often the primary cause of the problem. If you put them next to a strange female, it compounds the problem. I've had hundreds of patients come to me after a very disappointing time with a surrogate. Even if they have seen her for a very long time, premature ejaculation patients are usually shy, confused, and sensitive. Impotent patients are often frightened, disillusioned, bitter, apathetic, and confused. Put either of them next to a strange female who obviously has no personal feelings for them, along with their special problems, and you've got a very strong prescription for failure.

CHAPTER

THREE

3
FEMALE SEXUAL DYSFUNCTIONS

IN MY PRACTICE, I have seen very few women who suffer with sexual dysfunction. I believe this is partially due to the fact that women are permitted to express their emotions. Our society says it's okay to show anger and certainly grief. When life presents an injustice or a tragedy, her sex life is usually okay because she confronts the problem and releases it through tears. Of course, women do not need to become erect, but they do need to be lubricated. Vaginismus or frigidity can occur. There are many sexual inadequacies women can get, but they don't get them very often. They can be treated in much the same way as the male sexual dysfunction excepting that the cymatic stimulator is applied to the glans clitoris. The percentage of women who suffer from a psychological sexual impediment is next to nothing. I do not treat a purely physical condition and will refer to a gynecologist.

Inorgasmia

Some women have never achieved orgasm. Female sexual dysfunction has been referred to as frigidity. The woman with low desire and/or no interest in sex will include an inability to fantasize during masturbation. Inhibited females report an absence of feelings, even irritation, anger, tension, anxiety or disgust (Kaplan, 1979). She may become phobic when it comes to sex. The female orgasm is described by many doctors and lay people as rhythmic contractions of the vagina. Masters and Johnson have defined it as facial grimacing, generalized myotonia, carpopedal spasms and contractions of the gluteal or abdominal muscles. The dictionary defines it as a complex series of reflexes at the culmination of a sexual act. Inorgasmia: an impairment of the desire and excitement phases.

Systematic Desensitization
(Treatment for Inorgasmia)

When anxiety plays a central role in dysfunction, systematic desensitization is a promising treatment. The client is trained to relax their muscles. Anxiety provoking stimuli is generated. In their imagination, the client is made to face feared situations. The client is then instructed to engage in real-life sexual activities to reduce or increase the anxiety evoking properties. In attempting to let go of "inhibiting events," sexual arousal proceeds unimpaired. Improvement is due to spontaneous change in sexual functioning or change in other life areas, or an experimenter-derived measure of sexual behavior and attitude.

Positive changes were reported on measures of pleasure, satisfaction and decreases of anxiety during sex. Investigation provides evidence that systematic desensitization treatment is correlated with positive changes in sexual functioning (Wolpe, 1958). Although this treatment appeared to produce changes in attitudinal and behavioral indexes, an insignificant increase in orgasm frequency was noted. When anxiety was reduced, the frequency of orgasm increased. For non-orgasmic women, anxiety reduction may only be a first step and additional treatment strategies may be necessary to produce an orgasm.

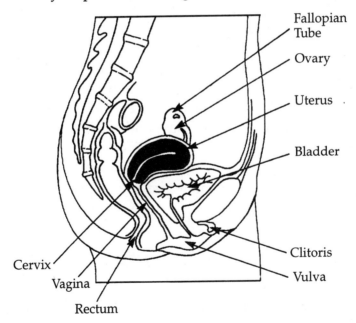

The Female Sexual Organ

Sensate Focus

The *Sensate Focus* technique was developed by Masters and Johnson (1970) in an effort to change behavior and communication patterns to remediate sexual dysfunction. The program offered to couples presents body touching exercises. This includes non-demand genital touching by the partner, female guidance of her partner's hands to sensitive genital areas, penile stimulation with direction by the woman, coital positions maximizing stimulation potential, and stop/start pelvic thrusting to enhance the female's pleasure (Herman & La Piccolo). Significant changes indicated that the entire female sample of primary and secondary subjects reported less sexual dissatisfaction and greater marital happiness (Herman & La Piccolo). This was after using home assignments, cognitive behavioral techniques, communication training, and systems conceptualizing (Loche & Wallace, 1959). These two doctors, Loche and Wallace, reported significantly longer duration of foreplay and intercourse, with greater frequency of orgasm, during masturbation and manual stimulation by the partner with intercourse. The original primary, non-orgasmic, reported experiencing orgasm.

(Carney & Bancroft & Matthews, 1978)

Women with the lowest levels of anxiety would respond favorably when receiving testosterone. Women with the highest levels, would respond best with anxiety-reducing drugs such as Diazepan. Sensate focus techniques have been the subject of a few empirical investigations. At present, they have only received correlational support for their effectiveness with primary, non-orgasmic women. Greatest gains were achieved by the Master and Johnson therapists. At present, sensate focus exists as a clinical technique without ties to other treatments or theoretical notions of the female response or sexuality. Study of the mechanism of treatment effectiveness or conditions of success for client or therapist variables has not begun.

Directed Masturbation

Masturbation exercises were conducted by La Piccolo and Lobitz (1972). These exercises begin with the patients examining their bodies and genitals with a mirror. They are asked to identify each part of their bodies in the privacy of their own homes. They then are to touch themselves to discover which parts give them pleasure. Then, mutual stimulation is given to the genitals until something happens or she becomes tired. If she does not experience orgasm, she is asked to use a vibrator. Her partner may join her. Since masturbation usually produces orgasm, Kinsey reported (1953) that the average woman reached orgasm 95% during masturbation and 75% from coital intercourse. Within fifteen minutes, all women experience orgasm with masturbation. Heinrich (1976) and McMullan & Rosen (1979) documented that the treatment, whether offered in groups or bibliography format, resulted in samples obtaining orgasms at least 47%, and often 100% of the time. Group comparison of factorial designs have indicated, that the absence of a partner, appears to have no detrimental effect on a woman's achievement of her sexual goals.

Hypnotherapy

When the patient is relaxed, and verbal suggestions such as "please yourself for pure enjoyment" are given, the estimated successful orgasm was 75% (Fabbi, 1976). While under trance, the patient may relate adolescent or homosexual desires accompanied by guilt when her parents discovered this. After three sessions, the patient reported a frequency of orgasm.

Why Female Sexual Disinterest?

Women have been conditioned to fake orgasm to please their partners. If men lost interest in sex and became interested in bulging waistlines or receding hairlines, much of the sex would be ceased. Women are taught from birth to be pretty at all expense. To win a man, we must compete with other females in the battle of "who's the prettiest." So, women spend their lives dieting, removing unwanted hair, fixing hair, having their faces and bodies cut up in plastic surgery, and so on, in order to win that man. Most women are never satisfied for fear he will find a prettier woman. This creates much anxiety and gives women little time for sex. Often, when women do find the time, they don't wish to have sex because they don't enjoy it. Men always enjoy sex. The causes of many female sexual problems are not always sexual. The sexual scripts women are handed are: "Nice girls don't," "Don't touch yourself down there," "Romance is all," "Sex equals intercourse," and "The modern woman does such and such," etc.

Women receive different cultural messages about sex than men. These messages are confusing and sometimes impossible to follow. Being pretty is the first script we are handed. The second is being thin. The third is being popular, and the fourth is being sexy, the list goes on . . . For women food is the single most reliable source of pleasure, providing much more powerful gratification than sex. Most female sexual problems stem from poor self esteem, lack of mutuality, non-physical intimacy in relationships, and juggling responsibilities in the home and the workplace. Many women feel, "If he only paid more attention to me out of bed, it might help how I feel when I'm in bed. He expects me to turn on in a second—as quickly as he does." Most women report very little sensitivity in the nipples or the vagina. It compares to the sensitivity in the testicles. Women are most often distracted from their own needs for concern over their partners' needs. The woman who fakes orgasm is sabotaging communication with her partner.

Physical Causes of Female Sexual Dysfunctions

Conditions which affect nerve supply to the pelvis and circulatory disorders are rare. Endocrine disorders are more common. Vaginismus is the most common and will be discussed in detail later in this chapter. Other problems are alcoholism, drugs, narcotic addiction, severe chronic disease, etc. Inhibited arousal is different from being sufficiently aroused and can come from shame, embarrassment, resentment, conflict, guilt, fear, tension, hostility, and depression. Inorgasmia, or the inability to achieve orgasm, will be discussed later in this chapter. Sexual fantasies are urged when inorgasmia is diagnosed. It spices up sex.

The male, who usually makes the decisions as to when to have sex, creates anger, resentment, and frustration to the female which causes her to disrupt her sexual enjoyment. Turning sex into a goal oriented activity or a chore is almost always joyless, unspontaneous and unsatisfactory. Repeated unpleasant experiences with sex may reduce the female's arousal. When arousal difficulties arise, it is important not to shut off one's erotic potential. You must remain an active, happy participant. If there is something about your lovemaking style that doesn't excite you, do something to change it. Inform your partner. Tell him/her what you want and need. Your partner doesn't always know. Touch more; feel more; think less. Don't be thinking about your physical weight or hair or whether they will think less of you. Think, but only sexual thoughts—wild thoughts—if it turns you on. Oh yes, fantasize. Remember, you can't go to jail for what you're thinking. So have fun! Have fun with sex. If a condom makes an erection go soft, make a game out of it. Don't take it so seriously. Again, just have fun. That's what sex is for. Play with different ways of exploring foreplay. Experiment with being any personality you wish—James Bond or Scarlett O'Hara. Make sex fun!

Vaginismus

Vaginismus is a female malady which causes the vagina to have involuntary spasms. Penetrating the vagina becomes extremely painful, sometimes impossible. It is not a disease nor a physical disability. It is an emotional condition which manifests itself in a physical response sometimes called dysparerenia. It may be caused by congenital deformity, infection, hormonal abnormalities, trauma, allergic reactions, tumors, and insufficient lubrication. Psychological causes include fear of intimacy, fear of dependence, lack of self love, lack of self esteem, lack of trust, and feeling unentitled to fulfilling one's needs. Approximately 20% of women in the Masters and Johnson Institute demonstrate a degree of vaginismus. In the U.S. the statistics are 20%. These patients originate from all fifty of the United States and from thirty other countries. Seven percent have a physical basis. Statistics outside of the United States: 30% in the United Kingdom (women between the ages of fifteen and sixty-four) are estimated to have vaginismus. There is a higher incidence in women between the ages of fifteen and twenty-four, representing the usual period in a woman's life when she first attempts intercourse. One study reveals that women of all ages are statistically more at risk of developing vaginismus, than they are of having to seek an abortion. One doctor specializing in sexual problems, estimates that vaginismus occurs in approximately five out of every one thousand marriages in Ireland. Another survey reveals that sixteen out of every one hundred women consulted at one birth control clinic, were suffering with vaginismus.

The signs indicating this condition (vaginismus), is when the muscles surrounding the vagina, contract when intercourse is anticipated. Inserting a tampon or a finger becomes impossible. It may have been caused by a painful or clumsy attempt at penetration, sexual abuse or rape, traumatic pelvic examination, expectation of painful intercourse, poorly handled episiotomy or herpes. The partner will always feel guilty about this dysfunction, for a man will feel it is his fault. At times it is. He must learn to be patient and gentle, as well as being certain the female is sufficiently lubricated. Vaginismus is not inherited, and a woman can conceive a child despite suffering from this ailment. It can be dangerous to give birth with vaginismus. It can be successfully treated through therapy, medication with rest and hypnosis. Some women will grit their teeth and make themselves available for their mates. Pain-free sex is impossible under these circumstances. A woman must see a doctor to treat this condition.

Inhibited Orgasm (Masters & Johnson)

Inhibited orgasm or inorgasmia is the inability to achieve orgasm. A woman may reach much excitability, but still not be able to release to orgasm. She may inhibit orgasm because she fears being too much in charge. Mother keeps speaking in the bedroom, "All men are animals. You cannot enjoy sex. It is for *his* pleasure only." This ominous message remains with the daughter and inhibits her orgasm. Father can be a major protagonist on the early scene of life. "His pleasure is uppermost and there is no life without a man leading the way." These rules of sexual morality are to keep a woman's appetite dulled. Some women avoid orgasm altogether the way one might resist good grades to promotion, or the exercise of authority. "Don't compete with men" is the message. These daughters may masturbate up to a certain point and then stop, as if to defend against the idea that anyone or any mechanism can take charge. Aggressive women often shun masturbation as a waste of time. They enjoy making the man feel so inadequate and clumsy, that he couldn't ever arouse her. They fear orgasm, as it might deplete them of control over their partners. Parents of these viragos often appear to be the nicest people in the world. They live out their parents' hostile fantasies and have gained that most precious elixir—their parents' approval. She may feel self-annihilated when getting too close to anyone. Often, they cannot identify with their parents, and this is essential/required if one is to be sexual. They often feel ugly, too fat, too thin, too stupid, etc., and this will inhibit orgasm. The French phrase for orgasm is "little death." They fear their souls will go out of them. They must be in control of themselves. Nature and Mom and Dad. They view orgasm as a mystery, and the journey might be a headlong passage into far away darkness.

Impaired Orgasm

Premature ejaculation in a woman is just as with men. Orgasm lasts a few seconds and is over, and she wants to go on to the next activity. She is often not certain she has orgasmed. The cause is often lack of sexual self-esteem and self-confidence. She fears asking for longer foreplay, varieties of stimulation, or wanting to please her partner by not asking for more work on his side. Unconscious fears may interfere with the fullness of a longer orgasm. Complete orgasm should take the entire body into broad areas of response for several minutes.

Orgasmic Anesthesia

Women who experience orgasmic anesthesia feel nothing at all, or feel a dull sensation resembling a slight tickle. Like the numbing effects of a potent drug, the distortions of such women's psyches deny them true feeling. Many women require long stimulation in order to orgasm—sometimes more than half an hour. When it arrives, it is unsatisfactory. These women feel abnormal. They feel they must conform to a male standard of brevity to reach orgasm. Delayed orgasm may occur when she feels doubtful of her partner's feelings. The more she tries to orgasm, the less effective it becomes. It's best to rest and try it at another time.

CHAPTER

FOUR

CONVENTIONAL
DIAGNOSIS AND
TREATMENT

There Is Help—Surrogates are NOT the Answer

THE MAJORITY OF MEN are helped with sexual dysfunction through scientific, sincere, compassionate psychotherapy from a trained professional, who will show them the way back to a loving, permanent relationship. The patient needs to feel worthwhile, inspired and loved. How can he feel accepted with the impersonal being of a surrogate? I've met many surrogates and all of them hated what they were doing. These ladies were often bitter, uneducated, promiscuous, and totally without compassion about their clients.

Case History:

Let me describe one patient who came to me for impotence with a large sack of pipes and planned to smoke them during the session. In my office, when you walk in, there is a nice big sign that states: "No Smoking—Do Not Smoke or You'll Be Asked to Leave." He needed treatment and knew that I was the only doctor who could guarantee a cure. He was sent to me by a urologist who had tested him for malignancy, hormone levels, and prostate problems, and a slew of other physical ailments. Incidentally, nicotine has been known to constrict the blood vessels which can lead to impotency. His medication might have seemed responsible for his impotency. When he came to me, he was desperate, but still he felt that he had to smoke. Of course, I wouldn't permit it. This left him with a pouting little boy attitude. Since he couldn't have his pacifier or nipple, he whined throughout the session. This was a well-known judge. He always had a pipe in his mouth except when he was working or sleeping. After the three-hour session, he thanked me for maybe adding a few years to his life; he felt good. But he said, "What will I do when I become anxious at work or around my wife?" I replied, "Why not carry a pencil with an eraser and suck on that? That's what you need to do; you need to suck. Also, you can stick chewing gum in your mouth—sugarless chewing gum—everytime you feel like smoking." He tried it and it worked. Three months later he called me and said that he had stopped smoking. Not that I specialize in this, but it's a hint. My five exercises made him capable of sustaining an erection for over thirty minutes at a time.

In the following chapter, I will explain why the caring partner is a must for a patient with a sexual dysfunction. The Masters and Johnson technique of masturbating and squeezing before ejaculating, is sometimes a very healthy and successful treatment for both premature ejaculation and impotence. Unfortunately, it does not always work because it is necessary to preface this treatment with Polarity. Polarity will be explained in detail in Chapter VII. Briefly, Polarity is a physical exercise where the patient learns to touch pressure sensitive contact points to unlock negative energy. These blockages are the root cause of disharmony in the body. Without this, the Masters and Johnson technique is only temporary.

For male impotence, several penal prostheses and injections have been introduced which have proven to be 30% effective, but have had numerous catastrophic consequences. Because we are cutting the penis or testicles to insert these items, there are many risks and complications. I would seriously consider the many other options before permitting a surgeon to enter this area.

Surrogate Therapy (For Males)

As I mentioned in the past chapter, I have never known surrogate therapy to work. Men with sexual dysfunctions are sensitive and shy. How can they go with a dispassionate stranger and expect to function when they cannot function with someone they care for?

Percentage Of Patients Cured Or Improved
(Controlled Experiment with Follow-Up)

January 1993	Premature Ejaculation	Impotence
Talking Therapy	50%	56%
Surrogate	4%	6%
Masters & Johnson	30%	——
Prosthesis	——	12%
Injections	——	20%
Drugs	——	30%
Beck Program	97%	98%

Note: This is based on three hundred (300) sexually dysfunctional patients studied.

Erection

The pelvic hypogastric and pudendal nerves gives rise to the cavernous nerves which supply the erectile tissue muscle of the penis causing an erection. This action of the central nervous system will induce penal erection. There have been several methods used in an attempt to give the patient an erection when he is unable to do so normally. Drug injection into the penis by Papaverine, Prosteglandine E-1 and Regagine have been used. The pitfalls, complications and risks are plentiful. The greatest danger is infection. The biggest pitfall is lack of success. Most drugs are not FDA approved. Yohimbe and Prosteglandine have been most successful. MD's are hesitant to use Proste-

glandine since it is not approved by the FDA and it
has been known to cause disastrous side effects.
Vacuum devices have been used and been reported to
cause bruising, discomfort and pain and not much
success as far as getting an erection. Penal implants
have been used which demand major surgery. Of
course, it cannot restore a normal erection. Most patients feel like mechanical men and come to me after
removal and then want holistic treatment.

Vitamins (For Males)

When the patient has been diagnosed as having a
low sperm count or a hormonal imbalance, I recom-
mend certain vitamins which can restore their normal
health. Daily: Vitamin E 600 I.U.'s, Vitamin C 4,000
Mgs., Vitamin A 15,000 I.U.'s, Vitamin B 50 Mgs. 3
times daily. Zinc 80 Mgs. daily, L-Angoline as directed
on label, Gerovatol as directed on label, or Zumba as
directed on label. Octacosnol: 2 capsules 2 times a
week, L-Tyrosine 500 twice daily on an empty stomach, Protcoytic enzymes (2 between meals), raw Orchic
as directed on label, Beta Carotene 15,000 I.U.'s a day,
Yohimbine as directed on label. *Do not take any of
these without consulting a nutritionist.*

A balanced diet is important. You cannot have a
healthy sex life if you consume large amounts of alcohol. Alcohol consumption decreases the production of
testosterone. Fats, fried foods, sugar, junk food, and
cigarettes should be avoided. Cigarettes cause the con-
striction of blood vessels and contribute to impotence.

Eat pumpkin seeds, bee pollen and royal jelly. Hot tubs and saunas have also led to reduced sperm count as does *excessive* exercise. In my experience, I have not had any conflict with patients taking drugs for diabetes, hypertension, tranquilizers or cocaine—even drugs for ulcers. Vasectomy for sterilization has been linked to prostate disease and cancer.

The testicles should be lower in temperature than the body to function properly. Tight pants or underwear undo the situation.

CHAPTER

FIVE

5

PREMATURE/RETARDED EJACULATION & IMPOTENCE DEFINED

FROM THE DIAGNOSTIC AND THERAPEU-
TIC points of view, it is easier and psychophysiologi-
cally more accurate to consider ejaculation and
impotence as a clinical entity separate from the classi-
cal concept of impotence. Retarded ejaculation is the
other end of premature ejaculation, and unlike the premature ejaculator, the man with retarded ejaculation is
often tired of sex. He has had it all. Many celebrities fall
into this category. They are offered all kinds of sex and
are calloused to all. They are satiated to the point of
fornicating even for days and not being able to reach
an orgasm, unless other methods such as sadomasochism or morbid sex is introduced into their activity.

Retarded Ejaculation

Many people who are on drugs have difficulty
ejaculating. This becomes a larger problem for the

female, as he feels she cannot satisfy him and she becomes vaginally quite sore. Medication often causes retarded ejaculation. The retarded ejaculator must have his medication monitored or decreased slowly in order to discover how little he can take and still ejaculate normally. With the sexually satiated retarded ejaculator, it takes a great deal of therapy to help him realize his unusual need for unconventional sex practices. He will follow the same exercises as the premature ejaculator, except for added analysis to determine his need for enormous amounts of sexual deviation in order to reach a climax.

Types of Premature Ejaculation

Premature ejaculation is the inability to last for fifteen minutes after entering the vagina or other orifices.

The workaholic is a common sufferer from premature ejaculation. His extreme compulsion about his responsibility leaves no time for play or relaxation. He is very little in touch with his emotions. There's no time—anything and everything else is more important. A typical premature ejaculating patient is one who (let's call him Bud), through compulsive hard work, establishes a very successful business and is the patriarch and benefactor of his large family. He enjoys the role, but feels a deep responsibility. He is married with four children and any spare moments he had were spent taking the children somewhere for *their* amusement, etc. He needed to hurry through sex with his wife as the family needed his time. He was never important enough to spend time with. At puberty, he rushed his masturbating. Bud prided himself with the fact that he could shower and shave, and masturbate in four and one-half minutes. There were more important things to do—take the children to school, to the zoo, to the circus—and on and on.... For this man, little or no time was spent communicating or fornicating. Since his wife never complained overtly, this negativity towards her continued until they came to me for therapy. It seems that not only didn't she say anything about his lack of interest in her sexual satisfaction, but she was relieved when he ejaculated since the act ended. Yes, she hated it and therapy showed her and him why. When this man came to me originally, he complained that his wife had not wanted sex with him for some time. It used to be fine, but now she always finds so many other things to do. Through psychoanalysis, I discovered that Bud had a basic fear of water. He feared any kind of water—ocean, lake, rain, bathtub or shower.

As a child, it seems that he was frightened by the sound of water going down the drain. Because of this acute fear, he only washed parts of his body with cold creme (face cream), not water. Obviously, he never really cleaned himself and his odor simply turned his wife off. Only in my office could she finally expose her pure repulsion of his odors, the dirt under his finger nails, his bad breath, or his odoriferous mustache. This man was a fine dresser, but when I asked him to remove his clothing to do some water therapy, his body odor was atrocious. His therapy consisted of placing him into and next to water, leaving him to wail like a child in a tub water, and after the initial shock had passed, his wife began to soothe him as his mother never had. We had to expose him to water as often as possible and then reassure him that he would not be lost "going down the drain." This expression was used by his mother towards his father whenever she wanted to remind him of *his* unworthiness. The son began to hear this phrase as an indication of his unworthiness. This negativity continued into his adulthood, appeared and disappeared sporadically throughout his marriage. After several months of the water therapy, he was able to confront and deal with his fear, eventually gaining control of it. Everything improved, especially their sex life.

Some of the causes of impotence or premature ejaculation are poor penal circulation or hormonal imbalance. The causes of premature ejaculation are most often psychological.

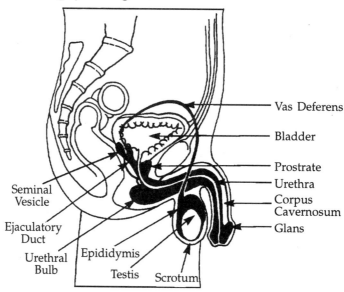

The Male Sexual Organ

EMOTIONAL EXPRESSION & AGE DISTRIBUTION
OF 177 PATIENTS AND FOLLOW-UP STUDY

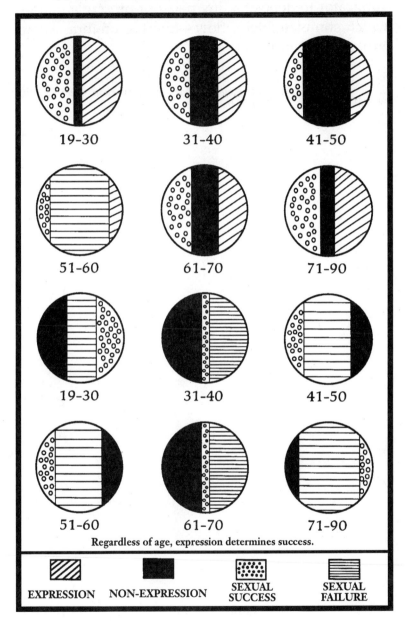

Regardless of age, expression determines success.

EXPRESSION   NON-EXPRESSION   SEXUAL SUCCESS   SEXUAL FAILURE

A new approach to *impotence:* men can use an applicator to deposit a tiny pellet of prostaglandins— which stimulate blood flow—into the urethra. About 65% succeed in having sex.

Sources—GOOD NEWS: *New England Journal of Medicine* (1.2.3)
BAD NEWS: *Journal of the American Medical Association, New England Journal of Medicine; Journal of the American Academy*

CHAPTER

SIX

6 THE MANIFESTATION OF S.E.S. (SUPPRESSED EMOTIONAL SYNDROME)

Cry—It Feels Good!

HAVE A GOOD CRY. YOUR survival depends on it. The heart bleeds and the tears flow and this is good. Psychiatrists and psychologists believe that crying relieves stress. The accumulation of feelings associated with whatever problem is causing the crying is released, and this allows for clear thinking and acting. Since crying is a healthy, natural form of expression, learning not to cry makes us unhealthy. And in terms of human development, less fit for survival. Emotions developed over a course of evolution as a way of communication, a way for a person to have his intentions understood and ensure his survival. As children we are scolded when we express our emotions. Johnny cries when he is frustrated, and as the years go by, he learns that control is what is expected of him and very early stops crying. After all, doesn't he want to please mama, then teachers, then peers? Doesn't he need to conform? They will call him a sissy! Doesn't he want to be masculine? What if he is injured? Doesn't he suppress his feelings? Now, let's look at Mary. She's thwarted, injured, humiliated. Isn't she encouraged to cry? Why does our society permit Mary to express her grief, hurt and stress, but discourage (even punish) Johnny for the same expression. Now look at Mary as she matured. She is rewarded for being feminine, which often means permission to express her feelings. Johnny matures with perfect control of every emotion. Isn't that what is macho or masculine? Since most psychiatrists and psychologists agree that the suppression of crying inhibits one's expression of fears, anxiety and anger, how can we expect him to be effective in his sexual performance when suppression of these emotions decrease the likelihood of survival. What do you suppose it does to his sexual ability? Is it healthy, normal, necessary to communicate feelings during times of stress? How we react to problems determines the amount of stress we then feel.

Means To An End

We usually react to problems by worrying about them. When worry begins to dominate our everyday life and we have no release, everything from colds to cancer, and from neurosis to psychosis manifests itself. When we have been trained not to express our feelings through tears, physical or mental activity, distraction, etc., there is no catharsis. This suppression becomes garbage to your body and soul, which soon develops into a severe anxiety and eventually a serious illness or dysfunction. And this is the syndrome I have been talking about, *Suppressed Emotional Syndrome (SES)*. What is the answer? Talk to somebody, take a walk, scream, go to a gym, jog, swim, go to a movie, read a book, have sex, or have a good cry. Try it. It feels good. True, when you are sad, depressed, sick, overwhelmed, etc., you are not always in a place where you can comfortably cry. But, you can always bring it up at home and cry. Focus on the incident. Think of only what caused the emotion and then let it go—cry. In my seventeen years of treating sexual dysfunction, I have had only four female patients. Why? They are permitted to express their feelings in our society and they do not have the suppressed emotional syndrome (SES).

Boy or Girl

Sexuality encompasses the biological, physiological, psychological, and sociological components of a person which is related to that person's gender. The concept of self as a sexual person is vulnerable to grief, as are our sexual roles, responses to sexual stimuli and sexual relationships. The early months of life are crucial to sexual identity development. The adult who molds a child's sexual identity by repeatedly telling the child through words and actions "you are a boy . . . do not cry," "you are a girl . . . cry," will eventually create a being who cannot function successfully in his (her) sex life. When associations between these highly significant early child experiences are revived in later life, the person may experience loss of secure sexual expression, plus a threat to his feelings of attachment for significant others. For some persons, the inevitable response to this compounded loss is to grieve. Now, Mary has been conditioned to cry—"it's okay, go ahead; girls do this." She substitutes no other activity. She does what is natural; she cries. Her subconscious harbors no garbage. She has expressed and released. Her sexual functioning is not impaired. And, it can be. There are numerous sexual dysfunctions women can get, but she does not get them because she expresses herself (Chapter III). Johnny, on the other hand, is conditioned to suppress his natural expression through control. He harbors all of the possible garbage. He cannot express his feelings. He will be subjugated if he does what his natural bent is—to cry. He develops premature ejaculation and impotence, not to mention all the other mental and physical repercussions he is heir to. Each waking moment becomes a struggle between threatened "I am Johnny, a person" and the conditioned mind (identity), "I am Johnny, a male." Where sexual performance is affected in the male, he feels severely threatened. Most men feel that the sun rises and sets on their penises. When sexual activity is halted or reduced due to trauma, the male will often feel inadequate about the rest of his life. This often precipitates regressive behavior affecting his masculinity. He often becomes helpless and childlike. Such behavior is typical by uncertainty in sexual identity and hesitation to accept responsibility. When intense, instant shock takes place, he may regress to childlike sexuality and have no erotic sensations, preferring cuddling and holding associated with mothering and care of a small child. The *prompt* anger response or grieving process drains energy from creative sexual expression.

All of this is treatable, curable, through the postponed physical expression of emotions. The patient must understand that his sexual dysfunction is real and legitimate, and that embarrassment or guilt is normal. All can be put into perspective and easily corrected in a short period of time. Many resources are available to help the patient minimize each of the sexual losses experienced. Sexual education, knowledge of the direction in the use of one's body for sexual expression, will provide a previously unexplored source of pleasure to a person experiencing a deficit in such experiences. While a patient may understand the lack of sexual energy and desires related to sudden trauma, he must learn that this energy can be directed to sexual success. Thus, crying, screaming, letting go as glorified in Bioenergetics (Chapter VII) must be learned.

Open discussion with patient and spouse on their joint sexual concerns will help them realize the myths of sexual dysfunctioning, and therefore, communicate and promote sexual health. Sincere communication with a patient promotes enhanced feelings about self as a sexual person, which promotes social and intimate relationships with significant others. Compassionate discussions will facilitate the subject's movement through the healing process.

Homosexuals

In my experience of treating male homosexuals, it holds with my theory that people who express their feelings do not have sexual dysfunctions. In seventeen years, I've only treated about forty homosexuals. Observe him, he expresses himself—he is sad, hurt, frustrated, he cries. Even in his draining years, he seldom suffers from impotence. He does not suppress his need to cry. He therefore rarely experiences premature ejaculation. I've treated several patients who had complained of premature ejaculation and impotence, and subsequently received therapy and became heterosexual. They had originally been homosexual. Invariably, in every case, they returned to a life conforming to the male image of suppressing emotions and eventually and sadly, developed premature ejaculation and impotence. Several heterosexual men turned homosexual. Surprising enough, they lost their sexual dysfunctioning. They began expressing. And how does the male homosexual develop? He may be conditioned by parents, etc., to react similarly to the female. Mom becomes his model. Perhaps it looks more interesting than whoever the male model is. Therefore he identifies with the female psyche. He sees no wrong in expressing his feelings. Most recently, there has been reason to believe that the homosexual is born this way or that a gene is responsible. The heterosexual male has denied his grieving for fear of the effect on his social or sexual life. Ironically, he has *caused* his dysfunction.

I have never had a female homosexual in my office. If my theory is correct, they do not have sexual dysfunctions, and they can't have them since society permits them to express themselves. There is no suppression and no emotional syndrome. Her psyche does not change if she reverts to a heterosexual life. Her personality may change. I've treated four male to female transsexuals who received injections of testosterone, had breast implants, and of course changed their genitals. They reacted similarly to the homosexuals mentioned above. Two of them stated that their orgasms were longer and stronger since the operation. Whereas, they had experienced all of their lives as homosexuals, their expression of emotions had not changed. Perhaps the added comfort and assurance that they were females relaxed them more and fuller orgasms followed.

Effects of Illness on Sexuality

Persons who suffer a severe illness for any length of time often experience lack of sexual desire or impotence. Certainly the depressive aspects of illness have the potential to lower the libido and may lead some people to feel unworthy of pleasure. When depressive aspects of grief deter sexual performance, some individuals will interpret their own reduced sexual interest as being a direct result of the illness and disability, and this may be true. My program of treatment will assure these people of successful sexual expression.

Case History:

A male patient of thirty-eight years old came to me in a wheelchair. He had a serious auto accident ten years hence and was unable to obtain an erection. My experience with spinal injury is limited, but most colleagues will agree that in many cases it can physically cause impotence. I refused to believe this in his case; he was sent to me by a highly regarded neurologist who stated that he was not neurologically impaired. The patient was suspicious, unyielding and understandably uncooperative. He had not lost his home, family or friends, but he could no longer work at his position as a sales representative. This devastated him. He was given a position with his company in a lesser degree, sitting at a desk part time with disability payments. The ability to provide for himself, his spouse and children, is closely associated with masculinity, even in today's more liberated society. Due to his initial inability to express his feelings, his added physical disability doubled his lack of resources and made treatment problematic, but not impossible. Mental exercises were simple, but he offered so much resistance it took four hours to treat him instead of the usual three. As usual, the Imagery was met with skepticism. But once completed, he agreed it relaxed him greatly. Since the Polarity was especially for this individual due to his injury, it was necessary to bring the blood supply to the genital area. With this simple exercise he reached his first semierection in ten years. When I showed him how simply he could do Polarity on his own, he was surprised and elated. The Cranial

Cymatics was applied for over an hour. Since he did not react normally at first, it eventually stimulated the erectile tissue in the penis. The reaction almost brought tears to his eyes, but he would not "let go" enough. His male upbringing dictated that his reaction must be controlled. I knew I would improve this with just the Bioenergetics. Unfortunately, he was not ready for this quite yet. His usual resistance told me this, so I proceeded with the Psychometry. There was so much fear in him due to earlier discourse with his father as a child. The alcoholic father beat him frequently. Guaranteed to cause SES. The added trauma to his spine demonstrated by rigidity and restricted patterns of behavior, made him extremely apprehensive. He needed to release all of the fear in his body. Fear is the number one cause of all sexual dysfunctions. Finally, we got to Bioenergetics. This presented a special problem for him as his movements were restricted and difficult. I adjusted this exercise to fit his particular needs. Even though he was beginning to trust me by this time, his negativity was most pronounced for the very physical exercises in Bioenergetics. Bringing about anger and grief, so difficult for most men, was extremely so for him. The second biggest cause of premature ejaculation and impotence is suppression of anger and grief. When I asked him to focus on an angry moment, he replied "I'm not angry." The usual reply. Most people have buried their anger so deeply it's hard to find. I asked him to imagine he was an actor and had to focus on an angry moment. After we achieved this, I asked him to beat the sofa with his fists, screaming anything he could think of. What he had harbored over the years, the rage he felt for his father, and the intense fury he experienced when told he would no longer walk, burst forth so fast I couldn't stop it. He so needed to express his rage at the drunken truck driver who hit his car. He had never expressed it. He had just said, "That's fate" and went on forgiving and trying to forget it (SES). But, this was not forgotten by the body. It reacts with sexual dysfunctioning. After doing another exercise where he twisted a large towel pretending it was his father and the truck driver's necks, he felt a wave of energy reach his mouth and ears—until his whole face was bright red, and he let out a huge wail and cried on my shoulder for a half hour. What a catharsis! I said, "Go ahead and cry. Women think men are sexy when they cry. It's true." He laughed and he cried for the first time in twenty years. He looked down at his penis and it was as erect as it had been when he was nineteen.

Bringing Forth Anger

Why did this happen? When an emotion such as anger is brought to the surface and expressed through physical and mental activity, endorphins are released (such as adrenaline) which affect the body and mind. The physical action of crying and laughing oxygenates the blood, causing it to rise and then fall. This intense action gives the mind and body the jolt it needs to bring the problem to the surface. When the problem is re-enacted and released, the body and mind are cleansed and relaxed. Norman Cousins, a great writer, wrote a book entitled *Anatomy of an Illness*. All the MD's agreed he would die in six months. He had an incurable, terminal illness. He made a fool of them by renting all the comedies he could find and laughing himself well. You can also cry yourself well.

Suppressing the expression of mourning, of angry outbursts or tears will eventually lead to depression and finally to mental and physical illness. When an injustice has been done, proper expression of anger is necessary. You are not encouraged to strike or injure the perpetrator at the time it occurs, but to express your rage any time later in your home, office or outdoors. It is best expressed in private since it involves antisocial behavior such as screaming, hitting objects and some profanity. This is explained in Chapter VII under Bioenergetics. If anger is present, grief is right under it. If we can experience the rage, we will eventually release tears. It's inevitable. When evoking anger from the patient, there must be a moment of total recall of an angry moment. Focusing in on this moment is imperative. No other thought should come into the patient's mind. The patient must circle in on all the events that occurred and exclude everything else from his mind. By emotionalizing the anger through thought, voice, and body, we then build the momentum until he is eventually screaming, hitting and letting go sufficiently, and recognizing his need to cry. This wave of released energy brings about a magnificent catharsis and the release of emotions too long repressed (SES). Letting his vulnerability erupt brings about a certain peace. At this point, sobs, shortness of breath, choking sensations, etc., become the tools of therapy. This experience, an emotional expression of anger and grief, are necessary for physical and mental emancipation.

Reaction to Loss

There is usually an immediate reaction to intense loss in most people accompanied by appropriate emotional expression. At the onset of loss, excessive activity, isolation, hostility, inhibition, apathy, depression, suicidal preoccupation, illnesses both mental and physical can ensue. Reactions to loss can be delayed, but should not be entirely suppressed. Suppression of anger and grief can be a predisposing factor in the onset and exacerbation of many types of sexual dysfunctions. This type of grief is characterized by feelings of helplessness and hopelessness. When this person sinks into a state of grieving, of giving up, there is a reduction in certain psychological and metabolic activities. This contributes to the emergence of premature ejaculation and impotence. It becomes necessary to express this helplessness as often as fifteen minutes per day for as long as the patient experiences the dysfunction.

When a child is exposed to loss, frustration, fear, or pain, he invariable reacts with tears, screaming, nervous laughter, and excessive activity. When he becomes an adult, this release is suppressed and all types of illnesses develop. When the male child shows excessive emotions, he is quickly stifled by the adults. This begins the onset of illness and the Suppression of Emotional Syndrome (SES).

While the female shares her anger and grief, the adult male turns aggression in on himself. His ego begins to lose its fervor, and acceptance of thwarted communication develops. The withdrawn reaction seems to be an adopted attempt to restore his inner equilibrium in response to the change in extreme milieu. A mechanism of conservation of energy is used until a new source of supply is available. Frustrations and renunciations lead to the growth of internal psychic structure and character deficiencies. The expressed anger and grief all replaced by internal punishments. When he expresses his feelings through Bioenergetics, he is ready to give up and act out his anger, and his punishments are deleted. Interference of this expression is most harmful. Helping the unmourned or the unjustified person to accept his pain, verbalize and emotionalize his guilt, is what we are talking about in this book. These feelings are vital to good mental and physical health. Now we see what a serious problem the over-controlled person can be. We are not discussing the seriously ill or the pathological person in this book. The deleterious effects of the over-controlled person cannot be overstated.

EXERCISES: ONE HOUR DAILY

Page 79   15 Minutes Masters & Johnson technique masturbation
(stop, pinch, start) culminating in ejaculation.

Page 80   15 Minutes Psychometry:
(1) Geometric patterns around flame
(2) Find anxiety . . . color/liquid . . . OUT!

Page 87   15 Minutes Polarity

Pages   15 Minutes Bioenergetics:
90–95   (p. 91) Legs, arms, head—100x
               (p. 93) Towel "GIVE IT TO ME"—25x
               (p. 93) Runner's stance "I WON'T FALL"—
                  25x
               (p. 92) Knees "I HATE"
               (p. 93) Punching bag
               (p. 92) Mirror "GET OFF MY BACK"—25x

CRY! CRY! CRY!

CRANIAL / TESTICULAR CYMATIC

Page 95   Two times a week.
               Two Cymatic Stimulators:
                  Place one at the back of the neck.
                  Place one just beneath the testicles.
                  (Be *sure* this area is DRY)

Page 76   (1) IMAGERY before sex. . . . Blackboard
                  Deeper
               (2) Foreplay
               (3) Enter
               (4) Sperm trip

I'm in control. No one else controls my erection or my ejaculation *but me.* Because I *know* this, I'm giving my partner and myself *so much more* pleasure.

CHAPTER

SEVEN

My Treatment

My TREATMENT IS A REVOLUTION-
ARY new program for curing, not improving, premature ejaculation and impotence in one three-hour
session. I've been immersed in research for twelve
years in order to discover this cure. It is now perfected, and all you have to do is give me three hours
of your time, and I guarantee a cure. I am the only
doctor who presently holds the unique knowledge,
experience and track record to offer a guarantee for
curing these two conditions. Unlike some doctors I
don't have you coming back, week after month after
year for treatment, and with disappointing results. I
literally *CURE YOU IN ABOUT THREE HOURS.*

During this time we do five mental and physical
exercises. Imagery is a mental exercise which you must
learn to do before you have sex, whereby you attack
the psyche and dispel the conflict of unresolved emotional imbalance. The other four exercises are Polarity,
Psychometry, Cranial Cymatics and Bioenergetics. I

cannot fully emphasize how vitally important it is for you to *never ever* do any of these exercises without the help of a trained professional. It may be harmful to your health. I'm going to describe them in the book, but you must never do them without a trained professional helping you.

Imagery

Imagery is the exercise you do before sex. It's a matter of relaxation which is the number one cause of all sexual dysfunctions. First, you must darken the room. Make *sure* the room is quiet and dark. Make sure you're comfortable—no tight shirts, no tight ties or tight jeans. Nothing tight is to be worn because when you wear something tight, it cuts off the energy flow and this is the worst thing for sex.

You cannot have good sex if you don't have good energy. So, now we're going to learn how to relax, by lying down in a quiet room on a nice quiet sofa or bed. Close your eyes and take two deep breaths. In with the positive air and out with the negative. In with the positive and out with all negative thoughts. Imagine now that you see a blackboard, and you're going to pick up a piece of chalk and draw a circle on the blackboard. To the right of the circle write the word *deeper*, and take your time. Beneath the word *deeper*, write the word seven times, making it smaller and smaller and smaller, until it becomes a dot. Go back to the figure on the left—that is the circle—place the number five (5) in the circle. Look at the word *deeper*,

deeper, deeper, deeper, deeper, deeper, deeper, deeper. Go back to the circle and erase the number five, and place the number four (4) in the circle, and then look at the word *deeper, deeper, deeper, deeper, deeper, deeper, deeper.* Now, go back to the circle, erase the number four and place the number three (3) there and go over to the word *deeper* and go *deeper, deeper, deeper*—seven times *deeper* until it's a dot. Go back to the circle, erase the number three (3). Replace it with the number two (2). Go back to the word *deeper.* Look at it—*deeper, deeper, deeper.* In that way, slowly, until it's so small, it's a dot. Go back to the circle and change the two to one (1). Go back to the *deeper, deeper, deeper, deeper, deeper, deeper, deeper.*

1—IMAGERY before sex	Blackboard	5	Deeper
2—	Foreplay		Deeper
3—	Enter		Deeper
4—	Sperm Trip		Deeper

●

I'm in control. No one else controls my erection or my ejaculation *but me*; because I *know* this, I'm giving my partner and myself *so much more* pleasure.

Sperm Trip

At this time, you should be sufficiently relaxed to start your foreplay. Imagery is necessary before you have sex, to relax you for dealing with both impotence and premature ejaculation. After you've done your Imagery, you are ready to enter (have sex). But as a person who's impotent, you might say, "How can I enter if I don't have an erection?"

After you've done the exercises, you're going to learn that you are *capable of entering* and you are *able to get a full erection*. The moment you enter, you're going to start a sperm trip; right after Imagery. Concentrate! During intercourse, you're going to concentrate on your testicles. Visualize them, visualize the sperm, and imagine that you have complete control of your sperm, and that you're going to take it through your blood stream, just like that! While you're having sex, imagine that you're taking the sperm right up through your stomach, all the way up through your chest, up to your neck, all the way up to your head, and back down again—down your neck, out your right shoulder, down your right arm to your right hand. And when you feel it in your fingertips, you're to say, "I'm in control. No one else controls my erection or my ejaculation but me and because I know this, I'm giving my partner and myself so much more pleasure." Imagine that you can take the sperm all the way up through the stomach. Now you have it in the fingertips. You're going to bring it back up the right arm, all the way back up the right arm, across the chest, down the left arm to the left hand. And when you get to the finger tips again, say: "I'm in control. No one else controls my erection or my ejaculation but me and because I know this, I'm giving my partner and myself so much more pleasure." Now feel the sperm coming all the way up your left hand. Imagine it coming all the way up your bloodstream, up your left arm, back to your chest, down your stomach, down your right thigh, your right calf and your right foot. And when you get it into your toes again, you're to say: "I am in control. No one else controls my erection or my ejaculation but me and because I know this, I'm giving my partner and myself so much more pleasure." And now feel the sperm coming all the way up your right foot, up your right calf, up your right thigh, across your stomach, down your left thigh to your left calf and your left foot. And when you feel it in your toes, again you're to say: "I am in control. No one else controls my erection or my ejaculation but me and because I know this, I'm giving my partner and myself so much more pleasure." Now feel the sperm coming all the way up your left foot, up your left calf, up your left thigh and back to your stomach. And when you get it here, you're to decide how many times you want to take your sperm on a trip through your body before you decide to ejaculate.

As long as you do this exercise, you will remain erect and you will not ejaculate *until you wish*. __This Works!!__ Now, slowly take two deep breaths again; taking in the good positive air and releasing *all* negative thoughts. In with the positive air and out with the negative. Breathe slowly as you're having sex. I've been told by patients that this exercise is so profound, that sometimes they don't have to do the others. It seems like nothing when you're doing it, but it works!

Masturbation

The patient needs to masturbate daily so that he trains himself to be in charge of his erection and ejaculation, both emotionally and physically. You (the patient) begin by grasping your penis in an up and down motion. When you feel that you are about to ejaculate, you stop and pinch your penis just below the head. It won't hurt to pinch it. You then wait a minute and begin again. Masturbate, pinch, stop, start, etc. This is the Masters and Johnson technique, but they wanted the patient to *squeeze,* not pinch. I ask you to *pinch,* to be sure and stop the ejaculation before you go too far. This action causes the muscles in the penis to strengthen and helps you control your ejaculation. Your goal is to last for fifteen minutes and culminate in ejaculation. You may only last a few minutes at first, but you will learn to last for fifteen minutes eventually.

Psychometry

The next exercise is called *Psychometry.* Psychometry is the exercise to rid the patient of fear. Fear is one of the biggest causes of sexual dysfunction. People are not aware that they have these fears, but they carry them all their lives nevertheless. It causes sexual dysfunctions. Psychometry is an exercise which should be done for fifteen minutes. These exercises should be executed one hour per day. The patient is placed in a very dark room and told to light a very small candle and look at the flame. You (the patient), close your eyes and open them, and close them and open them, and then close them and open them. You will concentrate on the small flame and begin to make squares around the flame going to your right. Make sure you're in a very relaxed position, with all the lights out except the candle light. See each line forming a square around the flame—one, two, three and four. Back again to your left—one, two, three and four. Now, we're going to imagine we're going to make a circle around the flame. Slowly, go to your left—one, two, forming a circle, three and four and back again to your right—one, two, three and four around the flame. Now, form a triangle around the flame—one, two and three. Now to your left—one, two and three. See each line forming a triangle. Relax your mind. Now, you're going to form a diamond around the flame. Go to the right with lines—one, two, three, four, five and six around the flame to your right. Now to the left—one, two, three, four. Begin to open your eyes wider and wider until you see your whole perspective. Just see the whole area in front of you and then slowly begin to squint, and make your eyes smaller and smaller. Close them until you can barely see the flame at all. Then slowly open your eyes gradually until they're wide open. Take in as much as you can. And slowly squeeze them until you are squinting and then open them as wide as you can. Slowly begin to close them and squint tightly again until you can barely see the flame, and then open them wider and wider and wider, and then close them back down again. Keep looking at the flame until you can barely see it and now close your eyes. Your eyelids are very heavy. You can't open them. They're so heavy. Close your eyes and rest your eyelids. Close your eyes and rest. Try to feel the pupils turning upward towards your forehead. Let your pupils roll to the back of your head slowly. Keep feeling them roll to the back of your head. Feel the pupils turning upwards towards your forehead. Roll them to the back of your head and now to the front. Totally let go. Let the pupils go and let them roll easily forward and then backward. Let them roll. Let the pupils do what they want. Let them relax—let them relax. It would be nice to have a little song in here if you can play a tape of a lullaby or any kind of beautiful soft music.

Now, I'd like you to imagine that you're a little baby peacefully sleeping. All your needs are taken care of. Whenever your needs arise, they're immediately fulfilled. You have nothing to worry about and you're being gently rocked to sleep. You feel satisfied, you feel so good, so warm and protected. Now, you're going to imagine that the time is going by and you're growing older. You are now fourteen years old. Imagine yourself lying on an air mattress in the middle of the ocean. Feel the gentle rocking of the water as you float across the waves. Mother is close by to protect you so don't worry about falling in. Feel the water move you calmly. How wonderful it feels. Mother's right beside you. How wonderful it feels. The breeze feels so nice. The sun feels so warm and comforting. It feels so peaceful. All is peace and quiet. All but an unwanted anxiety that was planted somewhere in your body. This anxiety has caused you to be dissatisfied with your life. This is you speaking from now on. This is you speaking, so be kind. "I want to go through my entire body and find the fear and dispel it. I'm going to find the anxiety and make it leave my body. The fear of sexual performance is causing me to feel inferior. I am not inferior—I am not inferior—I am not inferior. I am only defeated to the degree that I am willing to be intimidated. Again. I'm only defeated to the degree that I am willing to be intimidated. I alone have the power to succeed in my sexual life. I choose to succeed and grow—I choose to succeed and grow. I choose to find new levels of joy through a satisfying sex life—I choose to find new levels of joy through a satisfying sex life. I will practice my exercises religiously. I will practice them daily. For I know that problems are only situations for which I have not been trained. Repeat—*problems are only situations for which I have not been adequately trained.* I am training myself to find sexual fulfillment through the exercises Dr. Beck gives me. I will no longer suffer with sexual problems if I practice these exercises religiously. I will concentrate on relaxation, feel myself floating on an air mattress across the ocean, and feel the gentle rocking of the water as I peacefully rest my eyes and allow the pupils to roll to the back of my head and slowly to the front. Allow the pupils to roll to the back and then to the front. Just let go.

(The Patient speaks to himself and says):

"This unwanted anxiety that was planted into my body at a very young age is being reconditioned. I am going to go through my entire body and find this anxiety. Is it possible that this anxiety was planted in my head, my eyes, my nose, my mouth, my lips, my tongue. I am going to find this anxiety and get rid of it. I'm going to close my eyes and concentrate. Look at the left eyeball from the back, see the side of it, see the front of it. When I feel a little tickle or pain, I'm going to pin-point that sensation. Turn it into a color, then into a liquid and feel it leaving my body through the bloodstream. Feel it going down my face, down the neck and shoulders, through the bloodstream, down, all down my hand and through my fingers, and then gushing out of my fingers. Out, out, out, out. Then, I'm going to go to my right eye and do the same thing. And then to my nose. All the parts of my nose—every bone. To my mouth, my lips, my tongue, my neck. And now I feel so, so relieved. I feel much lighter." Then you go through other parts of the body. You'll not neglect the fact that you want to go to the heart and all the parts of the chest, the lungs, and all the parts of the stomach—doing the same exercises, concentrating on it, pinpointing it. When you feel any kind of tickle or sensation of any kind, we turn it into a color and then a liquid and feel it leaving the body. Let it go—let go. . . . "I'm a winner. I'm 100% responsible for my life. I feel lighter . . . so much lighter . . . so much lighter." Concentrate again on every part of your body. Feel so much lighter. "I will set short term goals for myself." You go through the entire body—the stomach, the legs, the thighs, and every part, and do the same exercise of having it gush out. From the waist up, you have it come out of the hands. From the waist down, you have it come out of the toes.

(The patient continues this self-dialogue):

"I will set short term goals for myself—day by day, week by week, month by month, year by year." Concentrate on the chest area—lungs, heart. See each organ. Turn it into a color, liquid. Then when you feel a tickle or sensation—feel the tickle or pain coming and the liquid gushing out of the fingers—all the way out, out, out, out. Say, "I feel so much lighter now . . . so much lighter. I will not be afraid to take one risk a week and make one change a month. Again—I will not be afraid to take one risk a week and make one change a month. Problems are only situations in which I have not trained myself to conquer. The more risks and changes I make, the more experienced I will become . . . the less afraid . . . the more satisfied . . . the more happy." Go down to the stomach area. Think of all the parts of the stomach and the genitals and the rectum. Do the same exercise. Concentrate on it. If you feel a tickle or a pain, pull it all the way out, feel it coming out your bloodstream, all the way down your thighs, all the way down your calves, all the way down to your feet, and gushing out of your toes—out, out, out, out. Focus, pin-point, see a color, let a liquid go all through your body and bring it all the way out—getting through it all the way—the calves, thighs, right foot, left foot. "I will let go of all anxiety. I will have no fear of failure for I am a winner. I will trust my body enough to let go of all the anxiety. I cannot give love if I do not love myself. I love myself. I love myself. I love myself." If you do not feel a tickle or pain, go on to the next area, but give yourself time to feel it.

Condition your psyche and say, "Thinking like a winner means not having to defeat someone else to reach my goal. It involves not demanding perfection. I will give myself permission to be the unique person I am. I can send away all thoughts of losing for I am a winner and a winner has choices. My choices lie in change and growth. I grow with changes. I grow with defeats. A consistent winner knows that you grow just as much with defeats as with victories. I grow with victories and I grow with defeats. I accept changes as I am an adult. I accept changes as I am an adult. (Keep repeating that). I accept changes with every new relationship. Remaining static is death. I choose life. I choose life. I choose growth. I choose life. I choose growth (Repeat this). I will never have performance anxiety again for I realize that I learn as much from failures as I do from successes. I know that worrying only inhibits performance. I will not worry about performance. When I have learned to do this through the exercises, I will be able to make peak performances out of all activities because I will be relaxed. I will be relaxed for I know that it is very important to be relaxed. I love my basic animal nature and am in awe of how beautifully my body functions. I respond to all its needs. I can see myself achieving and enjoying a long sexual encounter which will improve with every act. I feel the sensation and power with each sexual experience. I see this vividly for I know that I become what I think of most. I know that I become what I think of most. I am a winner 100%. Each day I will see myself achieving bigger, better, longer sexual goals, for I know that I become what I think of most. I will devote at least one hour a day envisioning my sexual expansion. What I get is what I set. I will achieve my goal. I will achieve my goal. Nothing will deter me. Nothing will deter me. I will be persistent and discipline my body and mind to function to my complete satisfaction. I will discover new horizons. I will overcome any and all anxieties to permit my body to perform to my satisfaction. I do things well because I'm that type of person. I will enjoy this because I'm grow-

ing. I will set no limitations on my desires and fantasies. I will never harm anyone. This is the only limita-
tion. What I respect I get. I respect my sexuality. What
I respect I get. I cannot attract the right partner if I do
not respect her. I get what I respect. I respect my sexuality. I get what I respect. I respect my sexuality."

Polarity

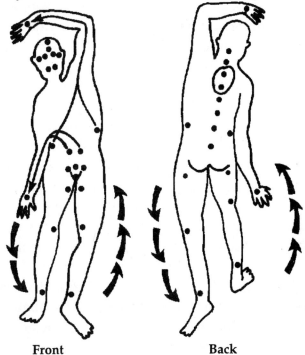

Front   **Back**

Pressure Sensitive Contact Points (Polarity)
Beginning from the back (get someone to help if
possible) press heavily on each point. Hold for the
count of seven; release and do the next one.
You can easily reach *all* points.
Continue to cover all points *DAILY*.

Polarity is a sensual physical exercise where I touch pressure sensitive contact points to unlock negative energy. These blockages are the root cause of disharmony in the body. This exercise brings the blood supply to the penis and causes an erection. A patient learns to do this for himself. It is quite simple. He simply has to touch certain meridians on his body and create an electrical current between the meridians in order to bring an adequate blood supply to the genital area. It's very simple and he should do it every day. It only takes a few minutes. Polarity is necessary for both premature ejaculation and impotence.

Place one finger at the base of the skull—at the cervical spine. Place the other finger just three fingers down on the vertebrae and alternate and vibrate to the count of seven with the fingers pressing heavily. Then, go down to the next three vertebrae. Alternate and vibrate to the count of seven, three fingers down. Alternate and vibrate to the count of seven, three fingers down, until you reach the bottom of the spine.

The next *Polarity* section will take you to the hips. Now we go to the hips and we place the fingers on the hips, the thumb on each hip. Alternate and vibrate to the count of seven. In order to bring an adequate blood supply to the genital area, you need to create an electric sensation between the meridians, which is what we are touching. Now you go to the pelvic bone. Sit down on your finger; find the pelvic bone jutting out. Put one finger there and one under the knee, under the tendon. Alternate and vibrate to the count of seven. Place one finger under the tendon of the knee on the outer part and the outer part of the ankle. Alternate and vibrate to the count of seven. Now go around to the other ankle, on the outer ankle underneath right next to the ankle bone, and then place one finger there and one finger on the knee. Alternate and vibrate to the count of seven. Bring two fingers (one on the knee and one on the hip) and alternate and vibrate to the count of seven.

Now that you've done the back, you're going to do the front. Begin on the forehead, placing one finger just where the hairline is and one finger beneath it. Alternate and vibrate to the count of seven, placing your elbows on a desk or table so that you have more pressure. The more you press, the better. Bring the two fingers under the eyebrows in the socket. Placing your thumbs there, alternate and vibrate to the count of seven to bring an electric current from one to the other. Go to the temples, placing a finger on either temple. Alternate and vibrate to the count of seven. Then bring the two thumbs down under the chin with the elbows on the desk and alternate and vibrate to the count of seven. Bring your fingers to each hip bone; to the *front* of the hip. Alternate and vibrate to the count of seven. Stroke the electrical current with one hand; all the way across your chest and down the opposite arm, all the way down to the fingers. Go between the middle finger and the ring finger (right between the knuckles). Press here—one, two, three, four, five, six and seven—alternating, that is. Go Down to the top of the pubic bone and press—one, two, three, four, five, six and seven. Then, bring down the electrical current, stroking your chest all the way across, bringing palm across, stroking your arm all the way down between the fingers of the other hand (between the ring and the big finger, between the knuckle). Count to seven. Go down to the inner thigh, then all the way up in that pelvic bone, pressing as hard as possible and the inner knee, alternating and vibrating to seven.

Alternating and vibrating the inner ankle, under the ankle bone. Shifting to the other ankle bone now, underneath, inner ankle, inner knee, inner knee, inner thigh. The last two meridians are just at the base of the penis, just beneath the testicle—lifting the testicle and counting seven.

Bioenergetics

We are going to now work on *Bioenergetics*. Bioenergetics is used to get the anger and grief out. Suppressed anger is the second major cause of premature ejaculation and impotence. We all have anger. It may have been there when you were two years old. You never dealt with it. It's still there causing your sexual dysfunction. Through Bioenergetics, we bring about a physical renewed activity and cause a spontaneous release of all negative energy. This provides a threshold for both active and passive roles in sexuality. What you have to do is make these exercises a way of life. *Forget what you learned about sex in the past.* Bioenergetics gives you a chance to express your anger so that you discover you will not be abandoned, punished or destroyed by voicing and acting out your feelings. The exercises for Bioenergetics are as follows.

Temper Tantrum

Lie down on a bed or sofa totally relaxed, preferably with no clothes on or just shorts, to begin. Concentrate on a time when you were angry and you didn't show it. I leave the patient alone for a few moments to concentrate and focus. You concentrate on it until you can remember a time when you were angry and said, "Oh, it's okay. It'll go away." It *won't* go away because the subconscious mind stores, and until you release it through Bioenergetics, you will have a sexual dysfunction. Concentrate on an angry moment. Start by raising one leg as high as you can get it, and banging it hard against the bed. And, raising the other leg and banging it hard. Then alternating legs—one, then the other, one, then the other—as hard as you can hit the mattress. Then, you start with your arms. Make a fist and clench your fists. Next, bang your arms hard against the bed and then coordinate the arms and legs like a temper tantrum. At this time, you're supposed to be as uninhibited as a boy of four. Imagine you're having a temper tantrum lying down. Start verbalizing. You can start by saying words like "No, I won't—leave me alone—No, I won't—leave me alone," or any words that come to you, be they in another language, as long as you are expressing your anger. Continue to do this until you feel some anger, and you will if you continue to do it and focus in on it, even if you have to do it one hundred times (that is fifty times with each leg).

Pounding While Kneeling

Another Bioenergetics exercise is to get on your knees, putting a pillow on the floor, and hitting the bed. Concentrating on a spot on the bed imagining that is your sexual dysfunction and hitting it as hard as you can, saying, "I hate you, I hate you." Keep saying this and say it fifty times (if you have to) until you feel the anger.

Lion

For the next exercise, sit and look at yourself in the mirror. See how you look when you're angry. People don't see themselves when they're angry. It's normal to get angry. We are supposed to suppress our anger. The patient is going to see how he looks when he is angry. When actors prepare for a role, they sometimes start out by imagining they are animals. So, you're going to imagine that you're an enraged lion; look in the mirror and watch yourself as you growl as loud as you can. Growl and *see* yourself growling as loud as you can. Looking in the mirror, imagine that you have a vampire on the back of your neck and it's sucking your life's blood. Now, imagine that you're getting it off by saying, "Get off my back, get off my back." And physically feel it coming off. Do this as many times as you need to do it to feel the anger. And when you feel it, let it go!

Towel Twisting

The patient takes a small towel and begins to wring it and imagine that the part he is wringing is the neck of his sexual dysfunction. And you're going to wring it and say, "I'm going to kill you. This is the end. I will never suffer again." Continue to do this until you feel that you've gotten it out of your system.

Take a towel. Twist it. Make the rolled part anyone or anything you like. Say, "I'm going to wring your neck. I'm going to strangle you. I'm going to kill you. I'll never have this sexual problem again." Make the rolled part of the towel your sexual dysfunction, if you wish.

Punching Bag

In this exercise, imagine you're hitting a punching bag. Hit it as hard as you can and continue to say, "I won't fail sexually. I won't fail sexually. No matter what the world does to me, I won't fail sexually."

I Won't Fall

Next, stand with your legs apart with your upper torso bent forward and fingers *almost* touching the floor... similar to a relay runner ready to go! Do not lean on *anything!* Say, "I won't fall, I won't fall, I won't fail, I won't fail sexually." Balance yourself and continue saying, "I won't fall, etc. . . . "

Tickle Me

Get yourself to be tickled by your girlfriend, your wife or whoever. Tickling is a very healthy exercise in order to release all kinds of endorphins. As I said before, Norman Cousins wrote a book called *Anatomy of an Illness,* in which he laughed himself well. He had a terminal illness and they all said he was going to die in six months. He bought all kinds of tapes to make himself laugh, and in doing so, he laughed himself well and he lived approximately sixteen years later and wrote another book. So, laughing is a very healthy activity. And, in order to have good sex, you must be able to be tickled and feel that sensation oxidizing the blood.

If you find it difficult to cry, and most men do, what you have to do again is concentrate on a time when you were very, very sad and you felt like crying (maybe when someone died or you saw a sad movie) and you suppressed it. Again, I say, do *not* suppress it! Try to teach yourself to experience the sadness and enjoy it. I mean *feel* it, express it, because if you suppress it, it causes problems. Try to imagine that you're a baby and that you want something desperately. If you can feel this—that you want it and that you're crying for it—and let your heart break. And after each one of these exercises (the ones that are for anger), if you do them hard enough and long enough you will eventually cry. This invariably happens. I had one fellow who was twisting the towel and turning it. And it's very hard to break a towel in half, but after showing his anger so much he actually broke the towel in half—tore it in half, which is hard to do—and he broke into great big tears because the emotions were so strong, and he said he felt so good. It was a tremendous catharsis. He hadn't cried since he was a little boy. So I'm saying cry, cry, cry. It's going to help your sex life.

Cranial Cymatics

After doing research with several neurologists over a period of seven years, we discovered that a vibrating stimulation to the supra-orbital nerve produces miraculous results. A physical vibratory stimulator is applied to the cervical spine. A spinal nerve controls the action of the ejaculatory duct. For premature ejaculation, it calms. For impotence, it stimulates the action of the erectile tissue. This whole exercise takes place in a supportive environment encouraging personal transformation. This treatment includes a careful integration of cognitive and experiential training tailored to the individual patient. When you have acclimated to the simple program, it will change your life. A catharsis must be brought about by increasing your capacity to experience heightened emotions, resolving characterological attitudes that have become structured in the body and interfere with its rhythmic and unitary movements. We learn to let go of anxious feelings in order to get control and approval.

Cranial Cymatics is done with the patient lying on his back. It should be applied two times a week, for about twenty minutes. A vibration is placed on the back of the neck (on that special nerve). This vibration calms the action of the ejaculatory duct for premature ejaculation and stimulates the action of the erectile tissue for impotence. This is a special vibrator given to the patient. No other vibrator will do. It really is something that has to be demonstrated. It cannot be done by a patient on his own.

Note: I emphatically stipulate that *NONE* of these exercises be done without the instructions and directions of a trained professional who knows and understands this treatment.

CHAPTER

EIGHT

"Good Grief"

"**G**OOD GRIEF" AS CHARLIE BROWN would say, is healing. "Good grief" is managed grief. Expressing grief with a supportive family and friends is necessary to good health. Of course, memories are forever. Scars will remain, but their outlines fade and heal with the exercise of Bioenergetics.

Mourning a Loved One

Express our grief and let it go. Emancipation and acceptance are inevitable when a clear thinking mind precedes it. When we lose a loved one, we must learn to let go of their physical presence. The grieving time is shortened when we embrace the sadness and feel its every rhythm. Begin with accepting the loss and letting your heart break. Talk about it with your friends and relatives and let them help you cry. How cathartic to let the tears cleanse until you feel your heart is bursting. Stay in the past for a half hour. Cry until you feel there are no tears left. Cry out to your lost child, husband, mother, father, or friend. Speak to them daily. Sure it hurts, but the outcome is eventual relief. True, only temporary, but crying it out daily will help to lessen the pain and help it fade. I didn't say erase it. It may be impossible to do this, but it is so good to feel the sorrow and let your heart break with big tears and after the crescendo, rest. Be patient with your friends. There's no way for them to imagine the pain that you're feeling. Most people think of death and all its aspects only when it touches them personally.

Most deaths are touched with guilt. No matter how far away or close (geographically) we were at the time of the death of a loved one, we always feel we could have done something to prevent or avoid it. No matter how much we did to help the person, we could have done more.

A strong religious belief can be a savior at the time of the death of a loved one. If the survivor believes strongly that their loved one is safe with God, this reassurance gives them the strength to see things a little easier. The pain is lessened when we feel that they are safe with God. Some people lose all their faith when a death occurs.

Case History:

Marion lost her son in an accident. She became catatonic and could not speak. She simply escaped from reality. She had to be fed intravenously. If she had only known about Bioenergetics and the natural expression of sadness, this reaction could have been avoided. She stayed in a coma for three months. When she became conscious of what had happened, she could still not accept her condition. She did not talk to anyone. She did not accept the death and she did not *cry*. She kept his room as he had left it and her heart almost stopped beating every time she passed that room. Her husband could not deal with the death, but he did have his daily work routine to distract him for a few hours a day. When he would return home, she would speak of their son as though he were still alive—discuss his homework, talk about what he ate for lunch. The husband was getting more and more insane from this. She would not leave the house, not speak to anyone, not answer the phone, and have all groceries and drugs delivered. She was a deeply religious woman, but this death left her totally believing that a God could be so cruel as to take this lovely child. What a cruel God. What could she learn from this? Hadn't she been a perfect mother? Guilt came deeply into the picture. Every little incident such as when she had scolded her son or had to leave him for an errand or a vacation became a guilt-ladened experience. "I should not have gone to Florida for that week. I should not have scolded him for not cleaning his room." These thoughts tortured her every day. Sleep was impossible without heavy barbiturates. Communication with her husband was becoming impossible. She developed an ulcer, and was again, hospitalized.

Male Grief

The father never expressed his sorrow. He tried to
be pleasant with his work associates and they tried to
cheer him. He recognized the reality of his son's death,
but never shed a tear. He neglected his health, not
eating properly, not sleeping, not talking to people to
express his anger and sorrow. A therapist would have
been in order here. He refused to go. All night long he
would repeat, "He's really dead. He's really dead. He's
really dead." There was no relief anywhere. He refused help. The son had disappeared. When the father
spoke to him on the phone and said, "If you think so
little of your parents that you'd rather be with someone you won't even share with us, then stay with her."
This statement was the last he made to his son. Three
days later the police reported his death. Dad felt that
he was the executioner. "If only I had demanded that
he come home." He was only fifteen years old. He
needed guidance and discipline. "I didn't do the right
thing. I'm guilty of causing his death." This self-pun-
ishment went on for three years, after which he had a
massive heart attack. His heart broke and he died. He
had always been in perfect health. Unfortunately, both
parents could have benefited from accepting and expressing their grief through Bioenergetics.

"Good Grief!"—What Is It?

Cry big tears and let go. Plan constructive grief,
concentrate on the sorrow, find a half hour to do noth-

ing else but cry. Think about the sorrow and nothing else. Cancel all other thoughts. Cry until you feel that your heart will burst open. When a death occurs, regardless of whether it was sudden or not, the survivor is often paralyzed. Tears will not come. It all seems so incredible. When the shock has subsided, the tears will come forth if the survivor gives a half hour a day to actual focusing on the loved one and letting the emotions come out.

Ways to Cope

When the Mother went to her priest and told him about the death, the priest said, "God must love you so much to make you suffer this way." She was horrified and angry. Was this all he could say to comfort her? She left her religion and joined a Buddhist friend. After joining this new religion she felt relieved after she came out of the hospital. Talking to others openly, chanting, expressing her sadness gave her the courage to go on. The unity she experienced as a Buddhist was necessary for her to heal. The open expression and the chanting practiced in this religion was an excellent outlet for her frustration. This religion eased her journey into this new world without her son. The journey can be less abrupt if we go gently and thoughtfully. She accepted work for the group which was beneficial for her when she needed to be busy. The religious change was beneficial to her. It does not apply to everyone.

Cry and You Cry Alone

It's a lonely world when we weep. Does it have to be? Not any more. There are bereavement groups in all of the hospitals. Most psychiatrists, psychologists, or therapists are familiar with their groups. See the list of places to go for help at the end of the book. It helps to be with others who are suffering as you are. In our culture most people do not like to talk about or hear about the dead. Neighbors and friends have a way of sweeping it under the carpet. "Don't think about yesterday, only the future." This is small consolation for someone who needs desperately to remember. Just when you need most to cry over memories, even laugh over them, you don't get to. You don't get the memories from your friends or your surroundings. People often treat the dead as ghosts never to be brought into daylight. They're reflecting their own weaknesses, however, not yours. People are afraid to talk about the dead as they fear their own deaths. They are also fearful that such words will make the grieving person more unhappy. So, it's up to you to set the stage for *"good grief."* Tell them that you're not disturbed, and in fact, you want to remember. They will never know unless you tell them. They will never relax unless you help them. What do most people say ... "I couldn't live without my child. How can you live without him?" How inconsiderate they are. Do you suppose they could ever understand just how you are hurting? Only people who have lived through it can understand. The most helpful thing they can say is, "What can I do to help?" To ask details of how the accident occurred or who was there or did he suffer and prolong the pain. To talk about his gifts, his strengths, his beauty, his intelligence, his warmth— these are things to discuss. Help them out. They don't know.

Let Me Help

At the funeral, her cousin made a worthwhile remark, "Be strong. Remember who you are." What a marvelous thing to say. It gave her such courage. Otherwise, no one spoke to her. It's as though they were afraid the bad luck would be contagious. She was silent in disbelief that her son was dead. Disbelief is the first reaction. There's a good reason why you relish any conversation about him; it assures you that he is not forgotten and that his life was not in vain. Tears from the depths begin about here, especially if the death was a sudden one. These are not tears of shock and dismay. These are tears of the soul. It's all right to cry. It's helpful. Cry out of bitterness. Cry out of the sense of loss. Cry just because it feels good. Best, for God's sake and yours, cry. It's part of being human. Be proud of your tears. You grieve much because you lost much. The love was real. A child grew up and died before he had a chance to live. This is worthy of heartfelt emotion. All that you are has been invested in one who is dead and you have a right to cry. So, go ahead and cry.

Guilt or Anger?

Sometimes it's hard to distinguish between guilt and anger when a death occurs. The survivor will always feel guilt and no one can stop this feeling. "If only..." is the usual remark. "I might have prevented his death if I had..." This is a natural reaction. It becomes morbid if it leads to neglecting one's health or grooming. And, sometimes it leads even to death. It's normal to feel some guilt. It's not normal to distort your feelings until you become ill. If it lingers and years go by, get help. You need to discuss this with a professional.

Anger is something else. We blame God and lash out at everyone for not understanding our pain. Anger should not be bottled up until it becomes destructive or until one turns to drugs, isolation, suicide or homicide. Before it gets to any of this, try Bioenergetics (Chapter VII). Release anger through exercises which vent deep feelings—scream, pound the bed, hit a punching bag—anything percussive. Patients will tell me they work out in gyms lifting weights or running or swimming. This is fine, but with this kind of bitter anger, the patient needs to hit something—be percussive with his (her) movement—talk it out with a therapist to place the anger where it belongs. Don't place it on your spouse, mother, children, friends, etc. Misplaced anger can be catastrophic. We're trained not to show anger in our culture. It's not polite. So, take a drive to the country, scream, cry, let go everyday. Crying and giving into being weak only helps oneself become strong. Sobbing loudly 'til we scream out at the injustice of it all. This is the healing this book wishes to impart. When the crying helps one feel cleansed like the feeling after a rain storm, this is the desired effect. The passion experienced here is what good grieving accomplishes (*The Healing Power of Grief* by Jack Silver Miller).

Why Grieve?

I agree with Dr. Worden, that to avoid grief is to avoid self love. Learn from it and remain vulnerable to love. We humans make strong emotional bonds and when these bonds are broken through death, the effects can be catastrophic. Sometimes these attachments come from a need for security and safety. When a child dies, this attachment is broken, since the security and safety has gone. The strong emotional anxiety experienced through the death of a child or a spouse is the highest form of stress imaginable. If the anger is released through crying, the mourner feels a catharsis and is relieved. Mourning is healing and must be practiced for equilibrium.

Face the Fact of Death

Grief cannot be started until the mourner faces the termination of life. Disbelief is the first reaction when a loved one dies. It can be helpful to strengthen the mourner's religion as the hope of being reunited with the lost one is comforting. Certainly, the funeral is beneficial when we see the person lifeless. I feel that the funeral is necessary to help one realize the finality of the situation.

Denial comes into action if we do not actually see the person inactive in the coffin. Some psychotic people actually keep the dead person in the house for days unable to let go until nature demands it. Cryonics is the process of freezing and storing a dead human body to prevent tissue decomposition, so that in the future the individual might be brought back to life when new medical cures have been developed. This system has not been proven successful to this date.

Seeking Help

Visiting a medium can bring us in touch with the person's spirit. It can be beneficial; it can be otherwise.

A Case History:

A lady who lost her daughter visited a psychic in desperation. She was a skeptic, but she felt, maybe there was some hope of reaching out to her lost child. The session with the psychic proved most interesting. This lady (let's call her Linda) was placed in a dark room with the psychic with only a Bible, a lit candle, and a glass of water. These are some of the props used by clairvoyants. As she chanted for the child to speak through the use of the water and a verse from the Bible, her voice became childlike, and it seemed to be the actual child's voice. Needless to say, the mother became almost paralyzed in disbelief. When the psychic said, "The child is holding up a poster. It is pink. It has musical notes on it and it has eagles on it," Linda was breathless, since she had been the artist who drew the picture. When her daughter was nine years old, she gave her a pink poster that she had made with the song "Happy Birthday," with coins glued into the notes. Mother had placed quarters into the quarter notes, half dollars into the half notes, and silver dollars into the whole notes (thus the eagles). Her daughter's spirit held up this poster. The daughter loved it and held it up for all the friends and relatives to see. Could it be that her little girl was out there and waiting for her to join her someday? Could this medium be authentic? Linda had never seen her before. They were from different states, thousands of miles apart. Linda had been very religious, but lost it when her daughter died. How could there be a God who let this happen? How could God be so cruel as to let this beautiful child die? Just an inkling of a possibility that her little girl might be there, somewhere, brought back her religious feeling . . . the hope of seeing her again. The medium revealed the knowledge that she knew exactly who was at the accident site. She knew the name on the little girl's toy that she slept with. It was a stuffed tiger and she called it "Tige." And the medium said this name. The mother did not cry. She left in a daze, too startled to do anything but take a cab home. She couldn't sleep and she couldn't cry. She became an insomniac for twenty years, never able to rest as she might miss the

sight of her daughter. Crying would have been her
salvation.

Work Through the Pain

Anything that allows the bereaved person to avoid
or suppress his pain can cause all kinds of illnesses.
Sometimes people hinder the process of mourning by
avoiding thoughts about the loved one. They use stop
thought procedures to keep themselves from feeling
the dysphoria associated with the loss. Idealizing the
dead and avoiding the reminders of them are other
ways in which people keep themselves from expressing the pain of grief. Relief comes from allowing themselves to indulge the pain, to feel it, and to know that
one day it will pass. One of the aims of grief counseling is to help facilitate people through this difficult
task of experiencing the pain of loss so that it is not
carried throughout their lives.

Mourning a divorce is also necessary. Isn't this a
death? Mourn this death as you would any other ces-
sation of a life you once knew.

Adjusting to a New Environment

The mourner begins to miss the things that were
done by the deceased such as sex, companionship,
accountant, gardener, baby minder, audience, bed
warmer, chauffeur, etc. Trying to take on all these tasks
or searching for someone to fulfill them is not recom-

mended for at least a year. After this time, they can adjust and react normally without pressure.

It is from frustration and a sense of helplessness that most people do not associate anger toward the deceased. It is vented toward the doctor, the hospital, the hospital staff, funeral director or clergy. This displaced anger may be turned inward and experienced as depression, guilt, or lowered self esteem. Suicide may also be a possibility. Many people will not admit their anger.

Case History:

Bill became a recluse when his father died. He was only fourteen years old, just coming into manhood. He refused to go to school, eat, talk, sleep, etc. This brought him to a therapist much to his refusal to go. When the therapist asked him if he was angry that his father died, Bill finally answered, "No, he had a heart attack. He couldn't help it." When questioned further, Bill became angry with the therapist and shouted that his father had left him all alone with a mother who was dependent and helpless. "Why didn't she go instead of him?" This really filled him with guilt and anger. His shouting indicated how truly angry he was. This was cathartic though. Shouting out the anger revealed his intense feelings of guilt. Maybe he asked his father for too much. Dad helped him with his homework, drove him to school, gave him all of the luxuries he could wish for. Maybe he was too hard on him. If he was not so demanding, perhaps his father would still be alive. Even after the MD explained that his condition was inevitable due to the deterioration of his heart, Bill felt responsible for his father's death. The anxiety and helplessness associated with his withdrawal left him paralyzed. The counselor's role is to get him to express his anger—verbally and physically through Bioenergetics (Chapter VII).

Sometimes sadness and crying need to be encouraged by the counselor. The bereaved is fearful of crying in front of anyone. This is especially true of boys. Even so, if we can get him to cry at home, this is an accomplishment. He must cry. The therapist may help in identifying the meaning of tears, especially if the mourner is not in touch with his/her feelings and the tears come slowly, if at all.

Time Will Heal These Wounds

Well, not entirely. There will always be a scar. The counselor can assist the bereaved to live without the deceased by facilitating emotional withdrawal from the lost one. Time must be given to grief (grieving). The bereaved must be shown normal behavior. The counselor must allow for individual differences in mourning and provide continuing support. As he progresses, he must examine defenses and coping styles and decide if they are healthy. If he spots trouble, he should refer the patient to the right source of help. If medication is called for, he should be referred to the doctor who can benefit his situation.

Grief counseling should begin as early as possible and be done in a comfortable, professional setting.

Mourning is Necessary to Detach From the Dead

The survivor's memories and hopes need to be faced and experienced so that he can reinvest in a new relationship. Fear often prevents this investment. "Suppose it happens again. I won't let it happen again. I'll live alone." Holding on to the past is even more painful and must be "let go." There are other people to be loved and we must move on to them. Learn from the suffering that you could love someone and can lose them. Mourning is completed when the bereaved person is able to think of the deceased without the painful wrenching quality of sadness and can reinvest his or her emotions back into life and the living.

With some people completion is never reached. Crying can help to relieve this tension. We feel that no one can fill the gap that is left when one dies. Yet, we must go on to care for our children, our spouses and our friends. We feel the pain will never end, but it does subside and pass.

Normal Grief Reactions

Defined as the more frequent behavior, sadness refers to the feeling of loss and is normally accompanied by crying. Anger, if not adequately acknowledged, can lead to complicated bereavement. Anger can come from frustration that there was nothing one could do to prevent the death. One tends to regress when an important person in one's life passes away. One feels helpless and anxious. Anger can be misdirected when it is not expressed through Bioenergetics or tears. Blaming others for the death—the doctor, family members, friends, God, anyone—or turning the anger inward toward themselves can be a risky maladjustment. They may become suicidal, drug dependent or accident prone.

Guilt and self reproach are common experiences of survivors. "If only . . . I had been kinder, taken the person to the hospital sooner, and so on . . ." It is usually a distortion, an irrationality and totally unrealistic. Anxiety comes from the fear that they cannot take care of themselves and won't survive. Then comes the fear of their own mortality. This can develop into a phobia.

Loneliness is always expressed by the survivors. The death of a spouse is always followed by the need to be alone to feel the presence of the lost one. This can last for many years. But, "this too shall pass."

Feeling listless and apathetic is a normal grief response. It's as though you can't leave your home, your bed. Life has stopped or been put on hold. The feeling of helplessness is often expressed by widows. "How can I live without him?" Sudden death is followed by shock. Disbelief can paralyze the survivor. Pining for the lost person is nearly always experienced. A feeling of relief and emancipation can follow if the deceased was tyrannical, evil, unloving, or feared. When a person has been ill for a long period of time, relief is felt and this too is a normal reaction (Dr. Worden).

Some people feel numb or paralyzed by a death. It occurs on the report of a death. It probably occurs because there are so many feelings to deal with. To allow them all into consciousness would be over-whelming. This numbness is a protection from this flood of feelings.

Physical Sensations

The most commonly experienced sensations re-ported by the people seen in grief counseling are dry mouth, lack of energy, weakness in the muscles, tight-ness in the chest and throat, oversensitivity to noise, hollowness in the stomach, breathlessness, unusual-ness of surroundings and depression. When there is a sudden death, disbelief is the first reaction. We feel there is some mistake—that we are dreaming. We of-ten hope for this. Total confusion is another major emotion felt. We start out to go shopping and forget to bring the money. One widow went to visit her friend in Pennsylvania and wound up in Maryland. Totally absorbed in her loss, she was not watching the signs. Her preoccupation with her lost husband became an obsession and she could only think of how to recover this loss. A short time after death, there can be a sense of his presence. She may feel that the deceased person is still present and so it becomes an obsession. She may hallucinate and believe she sees and hears her husband. Because the widow feels she cannot live without her spouse, this triggers off very intense but normal feelings of sadness and anxiety.

Sleep Disturbances

The bereaved person may experience sleep disturbances in the early stages of loss. They can be temporary or last for years.

Case History:

After Julie lost her daughter she began taking barbiturates to sleep. She soon became immune to them and switched to other stronger doses until she needed a sleep clinic to return her to her circadian rhythm. Her husband Jim experienced dreadful nightmares and night terrors. He would awake a dozen times and scream, "She's gone. She's gone forever. She'll never come back." He was afraid of sleep as he knew he would have these awful dreams. He became an insomniac as well. In this case, a psychologist was helpful, but not for a long time. It took fifteen years to restore him to normal sleep. During this time, his work and social life was in disorder. Overeating and undereating are other reactions to a death.

Case History:

Marion was never heavier than one hundred and ten pounds. When she lost her husband she began what is known as frustration eating. After two years, she had gained an additional hundred pounds.

The newly bereaved may find themselves doing absent-minded things.

Case History:

Carol would start off for work at 3:00 a.m. only to return when she realized where she was and what time it was. She lost her sister in an auto accident.

Many people withdraw socially. They become sociopaths. Sara lost her mother. This was a symbiotic relationship and the loss was monumental to her. She began to have panic attacks and could no longer work or meet with her friends. They seemed to be forever mentioning her mother and she didn't want to hear another word about her mother.

Dreams of the Deceased

Dreams of the deceased are common.

Case History:

Ron would dream nightly about his dead fiancee. He suffered intense guilt since he was driving the car that the accident occurred in. Even though it was not Ron's fault, his dreams had him in court, in jail. He saw her sliding down a huge mountain that was made up of dead people. In the dream he is incarcerated and cannot save her. He awakes in a sweat, calling her name. A dream specialist helped Ron to actually re- live this dream where he actually escapes from jail and saves her. After reliving the experience in the dream a dozen times, it helped him overcome this anxiety.

Sometimes avoiding reminders of the deceased can be a solution toward the decreasing of pain.

Case History:

Margaret had so many pictures of her grandfather who she had lived with all of her life. When he died, she could no longer look at them. She destroyed all of them and also all of his belongings. To destroy the photos was wrong; to put them away would have been better. Later, she was sorry she had no pictures to remind her of the good times and his warm smile. Calling out to him with a comment about his kind- ness brought her sighs and tears. Tears have a won- derful healing power. When stress causes a chemical imbalance in the body, researchers believe that tears remove the toxic substance and help re-establish ho- meostasis. The chemical content of tears caused by emotional stress is different from tears secreted as a function of eye irritation. Tests are being done to see what type of catecholomines (mood altering chemi- cals produced by the brain) are present in emotional tears (Frey, 1980). Tears relieve emotional stress, but how they do this is still a question. Further research is needed on the deleterious effects of suppressed crying.

Inactivity to Overactivity

The bereaved can go from total inactivity to restless overactivity.

Case History:

Laura lost her husband in an accident. He left her with three teenaged children to raise. Fortunately, her mother stepped in to help, for Laura took to her bed for months. This inactivity became overactivity when her mother left. She would scrub the house until her body ached and be sleepless at night. Therapy finally brought her to normalcy.

Fear of the Dead

Visiting places and treasuring objects that belonged to the deceased is a common reaction of the bereaved.

Case History:

Tina wore the same dresses and size as her dead mother, but she refused to wear them, for fear that the clothes were too precious. Another reaction is to avoid the deceased person's belongings. One daughter felt they were contaminated, even though the mother died of natural causes. This daughter felt that the evil that befell her mother might rub off on her and she might die if she touched these belongings.

Anger and Depression

Depression is anger turned inward. If the anger is not overtly expressed by the bereaved, it becomes the pain within. Bioenergetics (Chapter VII) outlines the exercises and their many benefits. Depression and grief may seem different, but they should all be directed towards the end of releasing. The grieving process is determined by: 1) who the person was; 2) the nature of the attachment; 3) the mode of death; 4) personality variables; and 5) social variables. If the deceased was a father, mother, son, daughter or any very close relative, the grieving will be longer and stronger, especially if Dad took care of all finances, drove the children to school, took care of the gardening, etc., and supplied a lot of love. Every family member will be affected. On the other hand, if he was an alcoholic, beat his wife and children, and caused nothing but pain, his death is a relief. If a child dies in an accident, it most certainly changes the entire family. Parents will feel this every minute. Siblings, even if they were competitive, will react according to their level of attachment. If the deceased child was ill for a long period of time, the death may also be felt as a relief. Tears are still necessary for relief. Mourners can take comfort in the phrases, "Time heals all ills" and "This too shall pass."

In order to help the survivor to identify and express his feelings, the listener must understand that when someone we love dies, anger is a very common reaction. If one does not grieve or prolongs grieving endlessly, it can cause many problems. Denying sadness and refusing to grieve can be responsible for many illnesses. If the sadness is not released to expression, the person can suffer everything from neurosis to psychosis and from colds to cancer. Releasing by crying and screaming or interacting with others as in Bioenergetics is a necessity. Groups should be pushed by the therapist. There should likewise be a time when grief should no longer be expressed. To prolong crying and angry upheavals endlessly is to complicate life. After a year or two (at the most), the crying should be lessened. The upheaval stopped. Only through Bioenergetics can there continue to be healthy grieving.

CHAPTER

NINE

9

DYNAMITE SEX FROM PUBERTY TO SENILITY

WHAT DO YOU SUPPOSE HAPPENS the first time you have sex? It is usually with yourself or in masturbation (or self love), a normal way to enjoy yourself. If you are not fortunate enough to have privacy, there is some chance of being surprised by somebody in the family. If a female member like mother, sister, aunt or grandmother has possible access to your room, there's the usual danger of anxiety. Beginning at this time, the pre-pubescent male will often try to masturbate in a hurry for fear of this happening. This is often where premature ejaculation commences—fear of being caught in that awful act. This fear remains with you until something is done to stop it. The exercise on Psychometry in Chapter VII will teach you how to overcome this fear and abolish it forever.

Starting with a quiet room and the sound of a soothing voice or music, you are brought from a state of anxiety to a state of serenity. You must be taught to rediscover the beauty and pleasure of masturbation. If your religion forbids it, think: Would God want to take away your greatest joy? I don't think so. I want to show you that masturbation is a good thing. Lack of it has been shown to cause premature ejaculation. For example, when you have your first urine in the morning, it comes out fast and heavy, since there is so much pressure on the bladder. The semen is stored in the epididymis which is in the scrotum, and when it is full from lack of masturbation, the ejaculation is fast and full. When you release your semen daily (for most people anyway), the ejaculation will be slower—much slower. This is especially true of young men as they store more semen than older men.

Masturbation is important for older men, but the frequency is regulated by the age. From the age of 12–40 ejaculation should be daily. From 41–60 ejaculation should be 3 times a week. At 61–104 ejaculation should be weekly. It is important to mention that the health of the individual must be taken into consideration. In the case of older men, an erection isn't necessary in order to masturbate. Neither is ejaculation. Erection will occur if he follows Chapter VII. In order to masturbate you must fantasize. Think of anything that arouses you. No one knows what you're thinking, go ahead and enjoy.

Physical impotence is far more common than is supposed. As of the date of this publication, there are at least 31 million men in the U.S. alone that are partially or totally impotent. And only men that have *reported* their impotence. The male has a polster muscle in his penis. This muscle relies on masturbatory activity. Like any muscle, it needs exercise to keep it erect and firm. Every patient that comes to me for premature ejaculation or impotence is a patient who doesn't masturbate, or does so rarely.

If we believe what Freud said, and I'm prone to do so, we should know that the child reaches his first climax at the third to fifth year of his life. After a period of inhibition he starts again in puberty. For the pubescent male masturbation is healthy and necessary. It becomes unhealthy, even morbid, if he spends too much of his time in this activity. Surely once a day is O.K. When he reaches 18 or 20 he should be ready for the other sex. He should be practicing intercourse or some other form of release along with masturbation.

Masturbation should continue even after having intercourse. This is to keep the scrotum filled with new semen and to prevent premature ejaculation and impotence as I described above. If he continues to prefer masturbation, which is unusual, there must be some intervention. Therapy is needed to persuade him to seek pleasure with females. Narcissism may occur when masturbation becomes the only outlet. A parent should step in and freely discuss this condition without inhibition. The parent must make the young man feel that masturbation is normal, but stress that physical interaction with the opposite sex is equally pleasurable. The parent should be direct, honest and explicit. At puberty the boy must be told that it is O.K. to masturbate, but that later females should be part of his sex life.

Narcissistic choices are often seen in homosexuality. This can be avoided if there's explanation and understanding. We're still learning about homosexuality, and how it begins. It is possible that it is congenital, it is possible that it is learned. We're not sure.

The young boy's self-contentment and inaccessibility often make it difficult for the opposite sex to get close. Therefore it is imperative that the parent school him through his adolescence so that he does not become obsessed with masturbation.

He must find a place to meet girls and socialize. Eventually he will start feeling comfortable in the presence of females. The meeting can begin with convention and proceed to games. Many young people become acquainted with the opposite sex through games played with a sibling. It can be beneficial in later life to stimulate an otherwise dull marriage.

Fantasies

Carl, Diane and Marie were sisters and brother. Carl was my patient. He came to me after his marriage had turned sour after four years. He simply had no desire to be with his wife. What began as an exciting four years of satisfying sex became mechanical and forced. I realized that Carl was a strict Catholic and was unlikely to ever stray from his wife Jenny. They were a normal, healthy loving pair with two children. Jenny had also seemed uninterested when Carl suggested sex. It seemed like a chore all married people must do to appear, or feel, adjusted and normal.

They had not had intercourse for over eight months. He was depressed and needed to find a solution. His conscious mind prevented him from expressing how disillusioned he was, even to his doctor. His conditioning was directed towards struggling along despite the obvious dissatisfaction. He trembled as he fondled a gold crucifix around his neck. I had to delve into his subconscious. I proceeded to use hypnosis.

We began to speak about how he felt when he touched his wife. He said she was always so negligent about her cleanliness; something he had never mentioned in his conscious state. Jenny would have her friend over for dinner and leave the dishes in the sink all night. Her friend had dirty fingernails and Jenny always seemed to let her hair go too long unwashed. Of course he could never tell her this. How unfortunate for their marriage that there was no communication.

He recalled when he was a child of 8 and his sisters Diane and Marie would play games when their parents were out. He would be the doctor, Diane would be the nurse, and the younger sister Marie would be the patient. The ritual or the game was always the same. Diane would get a basin with water and some cotton and bring it to Carl. She would remove Marie's dress and her panties and proceed to cleanse her body with the wet cotton. Carl would have to examine her vagina and rectum and announce that all was well.

While he was talking I couldn't help but notice his erect penis through his trousers. He said he could not become erect with his wife for the last few months. What Carl needed was to role play with his wife. It might be a good thing to get her to take a bath or shower, a hint perhaps. He said it would never work since his wife was so conditioned to simple missionary position sex. He had once asked her to get on top and she was horrified and angry.

I asked Carl to bring his wife in to see me. She refused. I called her at home and after a lengthy conversation with her, she agreed to come to my office. When she arrived I was surprised at how young she was. Carl was 34 and Jenny was 23. I felt a game idea would be perfect since she was so child-like. I introduced the proposition of acting out a game or a fantasy with her husband. At first she was reluctant. After carefully explaining that it was an innocent game, and that anything husbands and wives do together is OK as long as they don't physically hurt each other, then it seemed acceptable to her. I told her to imagine she was a nurse. I even provided a costume which is often very beneficial in cases such as this. Carl was given a doctor's coat. I went into the bathroom and handed her a basin of water and a wad of cotton. She looked perplexed. I pulled Carl inside and told him to reenact the fantasy by having her cleanse her hands and face with the cotton. When they came out of the bathroom she seemed all giggly and excited. Then I sent them home, but only after I pulled Carl aside and told him to reenact the scene at home.

I instructed Carl to position Jenny as the patient. He cleansed her face and hands and genitals. This erotic ritual aroused the couple. They proceeded to dynamite sex. She was cleaner and delighted over the feeling of the wet cotton on her genitals. She also confessed she enjoyed playing the patient or the submissive role. Carl fulfilled his subconscious desire of

playing out his fantasy. Having learned of their success, I provided Carl with many other fantasies which
placed her in the submissive role. The wet cotton gave
him the opportunity of asking her to bathe before sex.
She didn't mind knowing what would follow. They
were so grateful and very happy.

Many marriage or relationships become humdrum
and need a boost. Fantasy is a useful alternative to
infidelity. When partners can't relax enough to play
out fantasies it is helpful to go to a show, a movie,
watch TV or read a book together and then talk about
the situation and gently suggest playing the parts. The
first time the suggestion is made, it may not appeal to
the partner. It could be due to conditioning, religious
training or shyness. Try and try again. Don't give up.
I've seen thousands of couples come to life with new
approaches to sex.

In order to prevent the premature ejaculator from
ejaculating too soon, the game should be played slowly,
fully dressed, before progressing to nudity. The experience should be drawn out as long as possible.

Case History: (Dominant Fantasy)

Mel and Patty were living together for 7 years.
They had experimented with group sex, oral sex, anal
sex, and multiple partners. They came to me ready to
break up. It seemed that they had done everything
together and they weren't happy sexually anymore.
They felt they needed new partners. After speaking to

them in detail, I realized this wasn't true. That was not what they really wanted. They were deeply in love. They needed to role play.

They worked together in a law office. He was a lawyer and she his legal secretary. They had grown children and therefore they had 100% privacy in the home and office. I spoke with them together and then separately. I suggested a dominant scene. They were sexually satisfied with each other but needed something new. Mel confessed he was very successful in his practice but he seemed bored with Patty; even impotent with her. I suggested a beautiful leather garter belt with matching bras, stockings and high heels, and that he should give them to her and ask to put them on. Patty put them on and Mel fell to his knees and kissed her shoes and her toes and her legs. He was so aroused being in an inferior position. In his practice he was always superior, in charge, and dominant. Patty was overcome with desire seeing him in an inferior position. When he tried to kiss her vagina she stopped him. She walked away, lit a cigarette and looked down at him in scorn. He had not been this erect since he first met her.

He crawled after her, begging her to let him kiss her there. She answered him with an emphatic "no." Patty then ordered him to "make me a drink and bring me that rope from the kitchen." It seemed to come to her so naturally, this dominant position. He obeyed and fell to his knees again. She tied his ankles together and then tied the rope around his testicles. She commanded him to crawl towards her. He could barely move and it was a long distance towards that goal.

When he arrived at her feet she smacked him hard on the face again and again and then on the buttocks. She then commanded him to crawl back to his slave corner. He did so. This went on for over an hour while he sustained an erection throughout. This ended in fantastic sex every time. She would sit on his penis, and whenever he was about to ejaculate she would stop, and smack his buttocks. She kept him from ejaculating until she orgasmed at least 9 times. They are still together acting out more and more dominant submissive fantasies.

Why can't more couples enjoy these fantasies? Why is it so difficult for the male to express to his partner how he feels? Why must he go to a professional to learn how to enjoy exciting sex? You can have dynamite sex every time. Just use your imagination and go for it!

Case History: (Female Fantasies)

Many wives have expressed their need for discipline. Ellen was a cardiologist. She worked long hours at a difficult and stressful job. The ball was always in her court. She had to make decisions which determined life or death. Her husband Jerry was in a less stressful position as an accountant. Jerry was a faithful, intelligent, witty, and loving man. She had no gripes. If only he weren't so polite while they were having sex! He would accidentally hold her breasts tightly and apologize. When he entered, he would apologize, fearful of having hurt her. When he performed oral sex on her, he would do it so lightly, she had trouble reaching an orgasm. She told me she would like Jerry to squeeze her breast, really squeeze it while they were having intercourse. She fantasized about him calling her names and putting her down during sex. She wanted him to shove his penis into her without foreplay, even rape her. She wanted him to suck her with his teeth until it hurt, a little. She wanted him to stop apologizing. I said, "Why don't you tell him what you want?" She replied, "How could I do it? He wouldn't respect me."

I feel that when two people really love each other, there is no greater gift to give each other than an exchange of fantasies. I feel it brings couples much closer together. I admit that it is difficult to convince them of this. But if you keep trying, it works. Be persistent! Exchange positions! Switch roles from dominant to submissive—sometimes he's dominant, sometimes she's dominant. After all, genitals are only appendages of the brain—use your imagination!

Masturbation Fantasies

In order to masturbate one must fantasize. What about? Patients will tell me that they can't masturbate. Sound incredible? I thought so too. If you are very inhibited, then fear will prevent masturbation. If parents, teachers or siblings punished or frightened you when you began masturbating, try to see a specialist in sexual dysfunctions. You might be traumatized by experiences rooted in childhood.

Case History:

Kim was a computer analyst, 37 years old, and obsessive-compulsive. She had two illegitimate children she had put up for adoption. She was fraught with guilt. Her alcoholic mother had warned her never to masturbate or she would be doomed to hell. Whenever Kim thought about masturbation, and she thought about it a lot, she remembered Mom's words and never dared to masturbate.

When she came to see me she was still unmarried. She was very nervous, a heavy smoker, and at least 70 pounds overweight. She said she never had an orgasm. Under hypnosis she revealed her hidden attraction to females. She remembered her first doll. She was five years old. She would take her to the attic, undress her, kiss her and actually chew her entire body until the doll was in pieces and there was nothing left. Her oral need was met, but her physical and mental release was never realized.

As an adult she had an unusual oral fixation which accounts for her heaviness and chain-smoking. At a later hypnotic confession she admitted playing with dolls when she was about 8, with her friends. She would take the baby bottle, which was shaped like a penis, put it between her legs, at the entrance of her vagina, and hump her friend in a masculine way. There was a lot of giggling. "It was all very innocent," she said, "After all, we had all our clothes on." She felt such guilt admitting this that she could never say this in a conscious state.

During her third hypnosis (age 13), she remembered finding her sister's soiled panties, wearing them over her face and smelling them. During her fourth hypnosis (age 14), Kim explained how she'd sit under the steps which led up to her house to peek up the skirts of little girls climbing up and down. During her fifth hypnosis (age 27), she said she would take the top position during intercourse with men but would never have an orgasm. She would stop right before she did. Fear, Fear, Fear. Her mother traumatized her into orgasm interruptus.

After therapy when Kim realized she would not go to hell if she masturbated, I proceeded to explain that she might be homosexual. She cringed at the word. Her religious, rigid upbringing would never allow her conscious mind to accept this. I patiently related to her the hypnotic revelations. After much hard work with bioenergetics, she began to realize she could be a lesbian. She even delighted at the thought after awhile. I'm happy to say Kim has a wonderful two-year relationship with her lover Kelly and has lost 70 pounds.

Case History:

Janet was fifty when she arrived at my office for treatment. She was very beautiful with a figure of an 18 year old. Nonetheless, she suffered from inorgasmia. She was a widow of ten years with two grandchildren. Her husband Jim was great looking, intelligent, gentle, generous, and witty. After ten years she still wept whenever she mentioned Jim. He was the only person that gave her love.

Her father died when she was 9. Her mother placed her in an orphanage. Why did her mother keep her brother and her sister home, and send her to an orphanage? She felt rejected and abandoned. Mother never visited, or even wrote, or even sent her an orange. She was all alone, always competing for Mom's love. Mother had always said that her sister was pretty and smart; not Janet. When Janet asked Mom why she sent her away and not her sister, Mom never replied.

When her father died, the family was poverty stricken. There was very little to eat and the two sons had to be fed, while the girls ate every other day. Janet loved her father and when he died she was heartbroken. Mother never hugged or kissed her. When Janet had her tonsils removed, Mother never showed an iota of sympathy or love. After all, her sister was the pretty one. Janet grew up to feel inferior; never quite good enough. She never knew unconditional love. She would be forever seeking her Mother's love and approval.

She remembered when she first got her period. She had no idea what it was. She put toilet paper in her panties. She overheard the boarder who slept in the big room with her sister talk about sex. She listened through the wall, catching patches of information. Her sister asked the boarder what syphilis was. The boarder said it was a terrible disease you get between your legs. Janet thought that must be what she had. She told her mother she had syphilis. Her mother was an immigrant and didn't know what that was. Mom told the uncle, who was a miserable sadist. He threw Janet across the room and she landed at the end of the table. She gushed blood. They brought her to a doctor who examined her and informed them that she was a virgin with her first period. No one explained what to do. The following day Janet went to school with toilet paper in her panties, and she bled all over her dress. The children laughed and ridiculed Janet. The English teacher took her into the bathroom, helped her wash her dress and showed her how to use a sanitary pad and explained what she had was normal.

This incident reminded Janet of the time in the orphanage when she needed a sanitary pad. She had to see a nun that was never available. When she got one pad she had to use it on both sides until it was beyond saturation. It was humiliating.

In the orphanage when Janet was 14, she was caught reading Boccaccio's *Decameron*. Sister Florence yelled, "This is pornography, you disgusting child." Janet was brought to a court of seven nuns that sent her to sister Mary Francis for punishment. Sister Mary Francis offered her potato chips, began to fondle her and then had sex with her. Janet was so startled having never heard of homosexuality. She threw out the potato chips and threw up all over Sister Mary Francis.

Janet became her plaything whenever Sister Mary Francis wished it. It remained a repulsive, dirty task until she ran away. She never told anyone—who would believe her? She had no family, nor had she anyone who cared.

On her 16th birthday, she borrowed 50 cents from Mother Superior, took a bus and never returned. She met an old man who took her in and fed her. She stayed with him and got a job as a waitress. Even though the old man was nice, she wanted her own place. She got a furnished room. She had a pleasant personality and got good tips.

After she paid her expenses, who do you suppose she sent the remainder of her money to? Her mother. Why? She had always tried to buy her mother's love. Of course her mother never returned love. Never, never.

Janet supported her mother. Though her siblings were all employed, she took on the entire burden. Mother had everything. Janet began working in the night clubs. She had a good voice and danced very well. She looked older and she had developed large breasts at an early age, so she passed for 18. No one ever questioned her. Anything she didn't need for herself went to Mother.

Janet married well and stayed with her loving husband until he died of a heart attack. She had a son and a daughter. Her married life was beautiful. Her only regret was that her husband had a job which took him away most of the time. She spent a lot of her time with the children. She continued her education in night school.

When her daughter turned 15 years old, she had an auto accident which killed her. This incident left Janet paralyzed. She needed to hire a maid since she could no longer move from her bed. When death comes unexpectedly to someone you love, especially your child, it can be devastating in its impact. When your child meets with a violent death, your sadness and rage at this senseless injustice is too great for your spirit to contain. You feel overwhelmingly guilty at not having been able to protect your child, however unrealistic such thoughts may be. You feel extremely vulnerable and powerless. The depth and chaos of your feelings may even convince you that you are going crazy.

The experience of grief after an unexpected death can be truly agonizing. So agonizing that you feel powerless to stop the flow. I told Janet that though it may not seem possible at the moment, we can work through her grief by moving beyond the trauma of the loss of your loved one, while still preserving the bond between you.

All of this was accomplished on the phone as Janet could not leave her house. She could not speak to anyone but me. I explained that losing a child often evokes emotions she has never felt before. I told her not to be afraid of her feelings. I said she may think that her intense emotions of sadness, anger, revenge, fear and loss of control are abnormal. But her emotions are a normal part of the acute grief associated with an unanticipated death.

She needed to bring closure to her relationship with her child. Thoughts and feelings she never fully shared filled her with a sense of incompleteness. Because this feeling of incompleteness continued, I suggested she write letters to her daughter, keep a personal journal, or speak directly to a photo of the little girl. This helped her cry.

When a death occurs, the survivors are often in such shock that they cannot cry. She said she would die from her broken heart. Crying removes chemicals such as manganese in tears. Releasing this chemical relieves stress. Her tears streamed down and she let them flow as freely as they would, making a pillow for her heart. It took a long time to get her to cry. She waited too long to express her despair. When people wait too long it causes a buildup of tension, often causing ulcers or colitis. Janet had waited too long to cry. She developed an ulcer.

After a year of therapy, her ulcer disappeared. She joined a support group which was beneficial to her, as she had been a recluse for so long. She found support, and in sharing in the darkness of despair, she realized that others were also suffering. She wasn't alone. Here she experienced a liberation of her emotions. She outwardly expressed her fear and agony. Such sharing not only eased the pain, but also rebuilt her trust, paving the way for loving relationships with others.

Grief is a process that does not end quickly or automatically or even predictably. The worst times usually are not at the moment the tragedy strikes. You are in a state of shock then. Often you slide into the pits for some months, approximately seven months after the event. Ironically, this is the time when people expect you to have gotten over your loss.

Janet is now living a happy normal life with a new husband because she followed my 3-hour program. I'm so happy for her. After all, life without love has no meaning. After years of crying therapy she could release and enjoy life and sex with a new, loving person.

In the above-mentioned case histories, the patients experienced traumatic events in their childhood, which later prevented them from functioning sexually as healthy adults. My 3-hour treatment program released the bottled-up emotions which threatened to choke off their virility and their ability to engage in healthy sexual activity.

CHAPTER

TEN

10 CULTURAL INFLUENCES & THEIR EFFECT ON SEXUALITY

W HEN MOST CULTURES ARE DEEPLY concerned about survival, the objection of the female to male sexual dysfunction is often nonexistent.

National Cultural Influences

Most authorities agree that few countries value the female. She is less than an animal unless she is wealthy (high-born). Money seems to be the great equalizer.

In America, the female may rise above her heritage.

Saudi-Arabia

Saudi-Arabian women have almost *no* privileges. If she is of royal family, she often keeps millions of dollars in the palace to avoid the risk of beheading, imprisonment (for life), arranged marriages, sexual abuse, violence, barbaric rituals, head shaving, female circumcision, sold as concubines, raped, even murdered by a husband. His reason need only be that he is tired of her. Her only defense is money!

A man can claim infidelity with no evidence. He can claim she murdered her baby with no body to show for it. The man is king. His word rules. She is not believed. Aren't all women liars? This is the attitude and almost no one can leave the country without money and lots of red tape, which is resolved with money.

The biological revolution began at birth. Jobs cannot earn what it would take to leave the country. Of course, the government would frustrate every effort. The woman is imprisoned, trapped, with little or no hope of gaining equality.

Morocco

The Nomadic woman is a slave to her husband or lover. If he provides for her, she must be prepared to have sex with him 5 or 6 times a day. She is defined by her man.

Europe

In France, if the man wishes to have his wife make love to another woman, she must consent.

German and Scandinavian men are more like Americans, in their regard for female equality.

China (1920–1995)

Open sex discussions among women are a rarity. Privacy is hard to come by as space is scarce and population is high. In Hong Kong premarital pregnancies are common. These women are in no position to expect a dowry or what is called a BRIDE PRICE. A woman in this position is disgraced. No longer are most women virgins when they marry. This is true even after corrections are made for changes in the age of marriages. Legalized abortion and sex education has increased. The marriage rate has fallen. The divorce rate has skyrocketed. The average age of first intercourse has fallen dramatically for both sexes, particularly for women, thus prostitution has diminished. Illegitimate births have soared. Pornography is rampant.

Semai

The Semai people are of very small physical stature and are naturally shy and non-violent. Their policy is of fleeing rather than fighting. The females are among the most unaggressive and retiring in the world. They are often under five feet. She cannot resist advances of the male. They call this "punam," meaning taboo. The woman feels she is denying him pleasure regardless of her needs. She does not wish to make his heart needy. It may affect his work and may place the entire village in jeopardy. She may be punished by the spirits with a hurricane. Any calamity can be blamed on her. There is no sexual jealousy and adultery is rampant. After all, she says, "I'm just on loan." The men will threaten them with death if they do not comply. These men do not worry about honor, productivity, or social born daring. Any illegitimate children are cared for by the country. The men threaten, but seldom do anything, since the women are easy. They refer to themselves as dumb, stupid slaves. Although such linguistic self-effacement may mask deeper feelings of resentment, we have no evidence of this and they deny any angry feelings. Children may not be disciplined. To put pressure on anyone is unacceptable. There are no competitive games or violence. There is no pressure for the boys to act strong or tough. There is no private ownership. "After all, we are only "LOANS" on this earth." There is no competition to earn more or crop more. There is no pressure to intensify or modernize their harvests. They are not competitive. They are cooperative. There is no need to prove one's manhood or strength. If a man wants some land; he need only ask. To refuse is "Punam."

India

If the Indian woman discovers that she has a female child in her womb she will try to have it aborted.

When female infants are born, they are often suffocated. This is due to the economy. It takes so much money to raise a female who will leave you and marry. The male will almost always support the family when he reaches maturity.

Amish

The male voice carries much more power (weight) than the female. This is a patriarchal hierarchy, which gives older men the most social clout and younger females the least. The final authority for moral and social life is rested in the male. Amish women are expected to submit to men at their will. The women tend to the children and all the chores in the house, and sometimes help with gardening, milking and many outdoor chores. They live a carefree life without electricity, refrigerators, can openers, dishwashers, blenders, mixers, microwaves, clothes dryers, irons, mowers, etc. Still, the male controls their sex lives—when he is through, they are through. They live in harmony, raising their children without these modern conveniences. The ban on electricity has eliminated all the machinery which makes their lives easier.

Trinidad

There is a great distinction between wife and mistress. They speak openly about sex and love.

Brazil

Passionate love is always a possibility, though not a necessarily articulated fact of life that customarily grew out of the sexual encounter.

Sweden/Denmark

Sexual permissiveness has deep roots. The medieval Norse culture was more permissive than the general European sexual cultures. It has long been a Scandinavian custom to regard dating rather than marriage as the starting gate for full sexual relations, even regarding marriage as an appropriate ceremony for welcoming baby. The relaxation of laws forbidding abortion began in Sweden and Denmark in the 1930s. Premarital virginity is no longer prized. The girls home is the place for teenage sex. Welfare benefits for unwed mothers are automatic. More than 45% of all births are illegitimate. Abortion is available on demand at an insignificant cost. Sex crimes are narrowly defined. Homosexuals are free to marry and adopt children. The divorce rate is high; birth rate is low. This freedom allows time for criticism on the part of the female towards the male. If he is dysfunctional, she is free to leave him, divorce, and she and her children will be cared for. Relationships are often brief and unfulfilling with all this freedom. It appears that whenever women are independent, there is a higher illegitimacy rate.

Germany & Eastern Europe

Similar to Scandinavia, with notable exception that an abortion is more difficult to obtain. In 1920 Berlin became the symbol of sexual freedom as much as N.Y., San Francisco, Amsterdam and Paris were in 1960.

Soviet Union

The Soviet Union made divorce available at demand. Under Stalin, severe punishment was instituted for homosexuality. Divorce was also difficult to obtain during his reign.

England

The sexual scene is much like greater Europe. The cult of virginity has vanished. Abortion and contraception are easily available. The illegitimate birth rate is high. Pornography is rampant. As in the rest of western Europe prostitution is not illegal.

Japan

The most powerful counter-example to the Puritanism-is-progressive conjecture is Japan. This country does not need the Christian ethic in order to prosper. The divorce and illegitimacy rates in Japan are low. These women are sexually conservative by American standards. This has nothing to do with sexual scruples. Few Japanese women are employed outside the home. Most of them remain dependent on their men for support. Could this be the reason for the low illegitimate births? Abortion, infanticide, contraception, male adultery, and pornography flourish here. Homosexuality is tolerated. Doesn't this parallel with ancient Greece?

Polygamy

Polygamy makes sense from the standpoint of protecting women. It increases the demand for women, thus a lower average age of marriage and a higher percentage of women who are married.

Being a nonexclusive wife withdraws one option from women. In most countries, the supply is much greater than the demand. Women in polygamous societies are conditioned to see the advantage of being a wife or being a spinster. The taboo against polygamy in most societies runs so deep, there is little need to justify it.

Case History:

While they were vacationing in the U.S., I met with a Mormon who had five wives. Polygamy leads to a patriarchal society! He came to me with four of them. It seems the others could not abide the fifth wife. She was deceitful, had very poor grooming habits, was completely uncooperative and was hated by all—*but* the husband.

He found her to be the most satisfying in the bedroom. Regardless of all her bad habits—she truly satisfied him. They were beside themselves. What could they do. He was reticent to explain why she was superior sexually. He was too shy, dignified and embarrassed.

When I questioned the man (separately) about her sexual activities, he said she was experimental. Not only would she perform fellatio expertly, but she would engage in role play. Asked to describe her skills, he said she was imaginative when she worked on him. She would skillfully embrace and lick his penis until it was hard. Then she would suck unrelentlessly until he could no longer hold back. She would make him wait for intercourse. Teasing, undressing slowly, sensuously, until he was crazy with desire. She was no more pretty than the others, just wiser to her man's needs. She would not simply open her legs to intercourse, she embellished the whole act. Sometimes she would tie a rope around his wrists and ankles and not allow him to touch her until he was mad with desire. Of course she wasn't a prostitute, although it may sound this way. She simply used her imagination realizing this was a man who had as much intercourse as he wished. She knew she had to offer him more.

When I explained to the four wives why he would not divorce her, regardless of their complaints, they understood. They left with the hope of pleasuring him in many new ways. The fifth wife would be kept apart from them if they could not tolerate her habits. One defense for Polygamy; it reduces promiscuity, by providing additional lawful outlets for male sexuality. Polygamy reduces the incidence of adultery and prostitution; increases the number of children. This husband wanted his wives to be happy. Most polygamous men are insensitive, brutal and treat their wives like slaves. Most polygamous marriages are tyrannical. The wives are *much* younger than he is. It is like a business; more managerial than caring. It is commercial and impersonal. Extramarital sex is almost impossible. The husband maintains surveillance over his wives and punishes them severely for merely flirting.

Female circumcision is practiced by many polygamous societies. Clitorectomy is a means of stopping infidelity or masturbation. Women have no rights when economically depressed.

CHAPTER

ELEVEN

11

AIDS TO
SEXUALITY

WOMEN WHO THINK THAT condoms
are safe are flirting with death. They are dangerously
misinformed about AIDS (Acquired Immune Defi-
ciency Syndrome). If they make sure their partners
use condoms; does this assure them they will be safe
from danger? NO! They may reduce the risk of AIDS
transmission, but they do not make sex safe.

A recent survey conducted by the Food and Drug
Administration in which more than 60,000 condoms
were analyzed, revealed that at least 30% failed to
meet the safety standards. There is no scientific proof
that condoms block the entry of the AIDS Virus during intercourse. There is a 10% failure rate in prevent-
ing pregnancy; the protection they provide against
AIDS is much lower, as the virus is many times smaller
than human sperm. Pregnancy can only occur a few
times a month—AIDS can be contracted any time she
has sex with an infected partner.

The likelihood of contracting the disease for women is low if she engages in oral sex. The virus must find an opening in a cut in the mouth. Anytime she engages in "wet sex," deep kissing; nipple sucking; she shouldn't come in contact with a man's semen. It is unlikely to be a danger, but even a hangnail could be a *possible* danger. It takes up to 6 months after exposure for AIDS to develop to antibodies detected by the AIDS test. Get tested a second time after the 6 months to be sure. Women should be suspicious of a man who is reluctant to get the test before they engage in sex. Men should be grateful and respect her wishes. Doesn't he feel safer with someone who is careful? Most heterosexual males are becoming cautious. "Our bodies are meant to bring forth life, let's not use it to bring forth death" (Dr. Helen Kaplan).

Other sexual activities can be substituted temporarily. Mutual masturbation, sensuous massage, vibrators and fantasies can be used. Rubbing against each other fully clothed is often gratifying.

Scientists are working on a vaccine to prevent AIDS, but it will be at the end of the 1990s until it is perfected.

Naturally the passion, romance or thrill is lessened, but don't take the chance of spending a lifetime in the hospital (or death) only to enjoy a moment of thrills.

College Students (American)

In 1974, The Human Communication Research Dept. reported College Student's risky behavior re-

garding AIDS. Two hundred and thirty-four (234)
undergraduates show that sensation seeking and the
sexual motions for a pleasurable relationship are directly related to all measures of sexual behavior (number of partners, incidence of unprotected sex and
percentage of condom use). The concern for health
and condom use is secondary to sensation seeking.
Sexual behavior is not generally inhibited by prevention (Psych Lab Journal Article).

Case History:

College students Phil and Mindy had been having
sex for 2 years. They were both Catholic—but not
religious. On their first date, they were more than
attracted to each other. She had had four men before
him. He had had seven women before her. They met
at a party in the bedroom waiting to use the bathroom. It was a long wait. They were both high. They
began kissing and he drew her to the bed where all
the coats were stacked. She was totally willing. He
was very aroused. Sure, they had had hundreds of
lectures and classes on the use of condoms. When the
moment arrived however they were too high and
excited to stop and put on a rubber. "It's so unroman-
tic," he said. "It breaks the spell" she said. Before you
could count 3—they were having intercourse.
    The next day they were both worried. Mindy was
not using birth control. Would she get pregnant? Phil
worried about AIDS. Subsequently; their relationship
blossomed into a trusting, monogamous pairing. Did

they think about a condom? Yes. They were both tested
and eventually had unprotected sex. The possible consequences were dangerous, but the thrill of the moment often takes over. Both 19 years of age; marriage
seemed uncomfortable. She is now using an IUD (In-
ter-Uterine Device). They came to me because Phil
complained that he could feel the IUD when he was
thrusting. They had experimented with condoms only
to have them break or slide off. Another month of
worry about pregnancy. Mindy tried the patch and
became nauseous. Phil said he was tired of getting his
penis sore from the IUD. They were arguing throughout the session. On subsequent sessions, there was
constant bickering about protection. They were con-
sidering breaking up. I try to keep couples together,
but I began to see a brick wall, as neither one would
concede. What could they do? I suggested oral sex or
mutual masturbation until and *if* they got married.
This worked for about six months, then they went
back to bickering. I suggested the patch again. Sometimes it takes a month or two for the body to acclimate itself. Mindy was cooperative. A few months later
they were ecstatic—all was well.

Case History: (Homosexuals)

Ron was 37, Tom 42. They met at a resort. They
had always been cautious and used protection. After
a swim, they were cleaning up when Tom snuck into
Ron's booth. Their passion was strong. Neither had a
condom in the shower—and you know what hap-

pened. They had been to an M.D. as Ron discovered he was HIV positive. Tom was bitter and angry as they had already been seeing each other for 3 months. Ron was sorry, he was in tears. Tom was pure anger. He had not tested positive—but he feared the next test. I asked them if they were using protection since they discovered the disease. "No!" they said. "I don't like condoms." Well, we know that *no one likes* condoms. Under hypnosis, Tom revealed his love of danger. Sex is no fun unless there are risks. I wondered how much fun it would be for him to suffer the pains of a dying mind and body. What about the pain of friends and relatives avoiding you?

Tom was a Masochist. He enjoys pain. Only a Masochist would continue having sex with an HIV Positive partner without protection. I looked for the causes of his Masochism and led him through a clear path of sensibility. Our work in Bioenergetics enabled him to see his wrongdoing. Their relationship has improved. They do use condoms now.

African-American Male (Journal of Black Studies 1994)

What did masculinity mean to a black man? This was asked of 32 black males ages 25 to 55. The generation and sorting of ideas were considered. What did being a man mean to them? The majority of younger men felt a need to have sex as often and with as many partners as possible. Did they use a condom if they were single? 90% said "No!" Older single men were emphatic about their views. "No glove, no love."

African-American Women (Women and Health 1994)

College women, 18 to 23, completed measures of body image attitudes. Positive satisfaction towards their fitness and health. Still, when that moment comes, they express a desire to complete the sex act and take the chance it won't happen to them. Even among the educated, when the moment of sexual arousal was so strong, the use of condoms were almost always neglected. There was very little differences in the less educated black women. The level of differences again come with age. The older black woman took pains to don a condom.

Case History: (Black Couple)

Karen (age 45), met John (age 30), in their office. They dated and after some time had elapsed, they became intimate. The evening she took him to her apartment, they became hot and heavy. He said, "I don't have a condom." She said, "I do." She took time to get it and put it on John in a very sensual way.

This could be the answer: Make a game of it. Make it fun! Make it sexy. If he loses his erection, don't worry; he'll get it back. He should care enough to want you both to be healthy.

HIV (Qualitative-Health Research 1994)

Twenty-three (23) persons were examined; ages 27–61. In the hospital or at home, there was the "LONER";

the "MEDIC"; the "TIMEKEEPER"; the "MYSTIC"; the "ACTIVIST"; and the "VICTIM." They can combine the Loner with Medic or the Timekeeper with the Victim, etc.

The Loner is often forced to be so. His medical condition may curtail social engagements. He may become the Medic; reading everything he can find on AIDS. The Activist spends his time writing, public speaking, demonstrating, etc., against the injustice; trying to activate the government to spend more money on research.

If the HIV sufferer is progressing into apathy, fever, headaches, and other AIDS symptoms, he becomes the Timekeeper and Victim. Persons living with HIV/AIDS inevitably deteriorate and previous efforts to exert control over the management and the course of their illness become futile. The Victim accepts the finality and relaxes.

Case History: (HIV Patient)

Paula was 32. When she was 13, she first stuck her arm with heroin. She had attempted to stop many times without success. Her school work was fair. She was sexually active and refused to use condoms. She always told her many partners she was infected. Most of them didn't care. After graduation, she had many jobs. None of them lasted longer than four months. Fortunately, she lived with her parents so that her expenses were minimal. Neither parent knew she was a heroin addict or that she was an HIV victim.

Paula nearly died at least 25 times. She was rushed to the hospital and the last three times the doctor told her mother. Her mother took it badly. She was already sick with a bad heart.

As her illness progressed, Paula had become weaker and suffered fever and headaches. She contracted Pneumonia and later Hepatitis B. No one visited her. Only her mother tolerated her daily outbursts of ongoing vomiting and bursts of angry, profane words. She hardly moved from her bed except to use the toilet. Her mother became her slave. When her illness finally became full-fledged AIDS, Paula became very thin, almost skeletal. She had no appetite, and she was apathetic. Later she had to be moved to a hospital where she survived on intravenous feedings. She became covered with ugly sores. Mom was always there. Paula passed away on her 32nd birthday. Mom passed away four days later. A wasted life!!

Drug Users

People who share needles when using drugs are subject to the risk of contracting the HIV virus.

Case History: (Sara: heroin addict–34 yrs.)

Sara began using heroin at age 17. Her mother expected a virtuous, religious daughter who realized that "Men only want one thing." Those words were repeated to Sara daily. Sara became afraid of men. She was suspicious of her father, her employer, her uncle, her mailman, the plumber, and so on.

Sara became a lesbian. Her need for heroin was insatiable. She felt she had it made. She thought she was happy. She had a lover; they were both drug addicts. She spoke about going to the lower west side of New York and pushing her arm into a hole to get her fix. The needle man never wanted to be identified. He remained anonymous by having his druggies push their arms into a hole for their fix. At the time no one knew about AIDS. Researchers believed in '94 (US NEWS Oct. 1994) there will *not* be a widespread breakout of AIDS within the heterosexual population. Recent research proved otherwise.

Recent Research: Medication for AIDS

AZT—Some advance in health but many side effects.

HPA 32—is a vaccine used only in Europe.

AIDS Symptoms: The earliest documented case of AIDS appeared in 1981, but doctors acknowledge that there were many unidentified cases in the 1970s. The virus may live for many days even in a dried inactive state, and then become infectious again. Symptoms include fatigue, fever, loss of appetite and weight, swollen lymph nodes, diarrhea, night sweats, skin disorders and enlarged liver and/or spleen. The first sight may be a tongue that is coated with white bumps. This is called candideasis which indicates a compromised immune system.

Para-suggested nutrients: (These can be purchased at health food stores.)

Aerobic 07; Dixyclor: Egg lecithin; Garlic tablets; Germanium; Protein supplement; Selenium; Superoxide dismutase SDD; B12; B6; Liver; Vitamin C; Bioflavonoids; Acidophilis; Coenzyme (CoQ 10); DMG; Kya-Green; Multimineral formula; Zinc; Copper; Protiolytic enzymes; Quercitin; Raw thymus; Vitamin A; Vitamin E; Aloe Vera; Fatty acids: L-Carmine; L-Cystine; L-methonaine; L-ornithine; RNA-DNA.

Important Note: None of the above should be taken before seeing a nutritionist.

Suggested Herbs: Silymarin; Cayenne; Echinea; Ginseng; Shittake; Mushroom extract; Parc d'auo; Red clover; Hpericin; Black radish; Dandelion; Chaparral; Bee propolis; Suma; Somastrin; Gingo biloba extract. Again, see your nutritionist first before consumption of these elements.

Always eat *fresh* fruits and vegetables which have been *scrubbed clean with soapy water.* Juicing is beneficial. Carrot and beet juice with garlic and onion added. Take Kyo-Green. Raw seeds and nuts; Grains. Rest and sunshine along with quality protein. Do not smoke, consume alcohol, caffeine, colas, sugars or red meat.

Determine which foods you are sensitive to. Drink steam-distilled water. The National Cancer Institute has conducted studies on a new anti-AIDS drug called ddL, which appears to be more effective in controlling the AIDS virus but less toxic than AZT. Unfortunately AZT eliminates more than the AIDS virus. It also leads to anemia. After 24 months the AIDS virus is resistant to AZT. Delavirdine is available to people with CD4 counts of 300 or less. At present, there is no cure. The best and only present answer is prevention with the help of condoms. *(See appendix for AIDS Health Organizations & Further Information).*

At the time of this publication Africa has been hit the most by AIDS; then San Francisco, then N.Y.

AIDS Hits African Soldiers

At least 65 percent of army-hospital beds in Uganda and Zaire are filled by soldiers with AIDS, and half the fighting forces in Zimbabwe and Angola are HIV-positive, says the World Health Organization.

"Combat killed 500,000 Angolans; the first years of peace may kill 1 million." WHO's Dr. Eben Moussi told AP. "Psychologically, physically, economically—Angola is not prepared for a disease that will hit with epidemic force.

Commentary

How sad that spontaneous sex is no longer possible or probable for uncommitted couples. Somehow

romance has become a thing of the past. Entering into
sex has now emerged into mechanics.

Newest Treatment—Protease Inhibitors. *Testing*—
Multidrug cocktails containing: IL2, A2T, TC3, Zerrit.

CHAPTER

TWELVE

12

No Kidding

RECENT INTERESTING FACTS ABOUT SEX involve some Egyptian wives who were brutally beaten as they refused to have sex with whomever the husband wished her to be with. In Trinidad, many transvestites stand in the streets begging for sex. No sentimentality is exchanged; simply an exchange of pleasure. In Morocco, Nomadic women will gratify their husbands anytime, and anywhere so long as he supplies them with a good household. They are defined by their men, only.

Women have more submission fantasies. Men have more dominant fantasies. Experienced people have more frequent and extreme fantasies than naive, inexperienced people.

"The One-Hour Orgasm," by Bob and Leah Schwartz, describes the many contractions which can occur during sex play. These contractions can be spread out and controlled until the woman is ready for an orgasm. By

delaying the orgasm, the sex-play can extend indefinitely.

During a Dominant Session, in which the dominant controls the scene, the orgasm is prolonged until the person in control is ready to permit ejaculation.

"Satisfying Your Partner Every Time" by Nora Hayden, is a recent book in which the author explains how the mate should massage the female genitalia with his penis to prepare her for intercourse.

Sex Surveys: (US News, Oct. '94)

Fidelity in America:
- 83% have had sex with one person.
- One half of Americans had only one partner in the past five years.
- Most Americans have sex six or seven times a month.
- Married people have the most sex.
- Percentage of gay men in America; 2.8 / gay women 1.4.
- Percentage of married couples of the same race 93%.
- Percentage of age difference within 5 years 78%.
- Similar education 82%.
- Wives who have had affairs 15%.
- Husbands who have had affairs 24.5%
- Percentage of men who have performance anxiety 17.1 / Women 11.5.

- Percentage of men who climax too early 48.5.
- Unable to obtain or sustain erection 20.4.
- Percentage of men whose beliefs guide their sexual behaviors 40.1 / women 47.6.

CHAPTER

THIRTEEN

13

SEXUAL FANTASIES/ DEVIATIONS

IF A FANTASY IS A free composition structured according to the composer's fancy—a deviation is an extreme abnormal sexual acting out of an unaccepted social or moral standard.

If two or more people are having sex and are fantasizing another person or situation, it can be totally accepted by society. If she is thinking about another man while they are having intercourse—this is a fantasy. If he asks her to spank him, this is a deviation.

In most fantasies, the people involved are *imagining* another stimulation.

Case History:

Marvin could not become aroused unless his partner humiliated him verbally and physically; kicked him and placed her shoe in his mouth. This is a deviation;

if you consider that the norm is sexual intercourse between two people of the opposite sex.

Of course, we could write a whole work on what is normal. What is normal for some people in other societies may not be normal for our society. We tend to try to follow whatever our immediate society practices.

Jane could not orgasm with her husband unless he growled like a tiger during sex. This is *acting out* a fantasy. She confessed that she fantasized that she was a piece of meat between the teeth of two hungry tigers. This, like most extreme fantasies can probably never be enacted.

Case History:

Betty and Paul were driving along when he accidentally hit a duck in the road. She said, "Let's stop and see if we can save it." As they walked up the hill, they saw another duck dragging the injured duck to the side of the road. Betty said, "Look, he's trying to save her." Paul said, "Look, again, he's taking advantage of her weakened state; he's having sex with her.

Betty confessed to me that she thought it was very exciting. She had been faking an orgasm with Paul for years, but since she witnessed this scene, all she had to do was visualize herself as the victim and Paul as the victimizer and she had multiple orgasms.

She was so worried about how abnormal this was. Remember, "You can't go to jail for what you're thinking." If some extreme memory or movie, story, etc., can bring you to orgasm—have fun with it! Don't ever feel guilty. It's only a fantasy.

Many famous people have been known to participate, not just fantasize, in extreme deviations. One very famous actor was known as the human ash tray. When he died, his body was discovered to be covered with huge cigarette, cigar burn marks from his unusual need for this type of deviation in order to become aroused. He was known to need outrageous deviations in order to become aroused. Hitler was known to need extreme deviations in order to erect.

Necrophilia

The power of fantasy can be magnified by the setting and certain props. A famous anchor woman confided that she loved the smell of death. She frequented many funerals without invitation. This is known as necrophilia.

The aura of death; the funeral surroundings all contribute to her need of unusual stimulation in order to orgasm. Normal sex bored her. "Romance is dull," she said. "I'm a morgue rat." She went so far as to become friends with a mortician. She paid him to put her next to a corpse. The smell of a newly dead person made her crazy. The attraction to blood . . . When she was on top of a cadaver, to purge blood out of its mouth, while rubbing her clitoris, was her incredible deviation. She wasn't hurting anyone and I warned her about AIDS. Burn victims or freshly embalmed corpses held no attraction. She would need this release at least once a week. She paid plenty for this. The mortician was not surprised—it seems it had happened before. The sex or age of the person didn't matter. They simply couldn't be obese. It seems that even to a necrophiliac, obesity is rejected.

Piercing

The youth of today are going in for piercing their bodies. It seems that since drugs are not as stylish as they were, something new has to come along to make them feel different—*alive* (as one young man told me). Unless he felt pain, he was dead.

It started at age 13 when he had several tattoos. This is painful. He had his unbelicus, eyebrows, lips, tongue and penis pierced. He would attach a long cord to a piercing on his body and pull 'til the skin broke. He related that the orgasm he experienced during masturbation was awesome. It was a higher ecstasy, beyond ecstasy. Without masturbation, he could never punish himself in this way. It was his way of escaping the human condition. Now at age 34, he could no longer enjoy masturbation without severe pain. Sex with another person was disgusting to him.

Today, the young people have put piercing into business. It has been practiced in many countries for thousands of years. It is done for pleasure. An erotic sensation is experienced when the piercing is done and after the rings are in place.

The Roman Centurions wore nipple rings as a sign of their virility and courage, and as a dress accessory for folding their capes. Victorian society girls wore them to enhance the size of their nipples.

Today, piercing is primarily sexual. When a dominant orders their submissive to pierce their body, the submissive derives great pleasure from suffering for their boss.

Piercing is a practice which should only be done by a professional, as infections could result. Some doctors will consent to do this. It is not for the faint of heart. Healing usually takes 6 to 8 weeks.

Royal Egyptians pierced their nipples. This was denied to commoners. The patient should have a well-shaped navel. A decorative stud or bangle is placed in the flap of skin above the opening. This directs the viewer's eye to the pelvic area.

The penis piercing known as APADRAVYA is described in the Kama Sutra as a modern day French Tickler used during intercourse to excite the woman. A common practice in India. It is piercing through the penis shaft behind the head or in the head.

A padlock through the Frenum was of European origin. The Frenum is located just below the head or crown of the penis. This piercing and placing of rings at this area was a form of guaranteeing chastity. The Frency cage device was to prevent copulation or masturbation. When the penis is flaccid, the ring becomes an erotic device to the man. He is prevented to copulate or masturbate, thus increasing his masochistic pleasure. Ring size is important. This piercing is not complicated and healing is rapid.

The pride of an Arab youth is achieved when he reaches manhood. The "Rite of Passage" is arranged

by his family and friends. One of the gifts will be a
silver stud or ring. There is a ceremonial piercing when
the stud is inserted through the left side of the scrotum between the testicles and the base of the penis.
They believed that this prevents the testicles from ever
returning to the groin which they describe in childhood. This gives visual evidence that the youth is now
a man. Wealthy Arabs install these rings/studs with
precious stones—the most precious being the Kuwait
Pearl. French Foreign Legionnaires have returned from
North Africa wearing this genital adornment on both
sides. It provides much stimulation when stroked.

In the South Pacific *"griche,"* pronounced "gresh,"
is very common among male natives. This is done at
puberty through the *raph perenis,* the ridge of skin
between the scrotum and the anus, at the inseam. After
piercing, a bangle is applied which enhances sensation and provides a convenient grip. The *quiche* is one
of the more erotic piercings. Pressure applied to the
ring greatly increases arousal and gentle tugging at
climax prolongs and intensifies orgasm.

Ancient Roman male slaves were often subjected
to the practice; some form of device being locked
through the perforated skin. With women, the device
was inserted through the labia. Both piercings are not
uncommon today, with no implication to chastity.

In Europe the genital ring is a symbol of betrothal.
The man has his fiancée's labia pierced and ringed
and she, his penis. Through their mutual pain a more
intense commitment has been made.

Women in the Sado-Masochistic scene often have
their clitoris pierced. It's a place to display jewelry.
Piercing must be done by an experienced piercer.

Prince Albert is a type of piercing in which the hole is put right through the head of the penis (horizontally). This is a popular style with many homosexuals. Urinating becomes a problem as the urine often tends to come out of the holes instead of the normal opening.

In order to safely and cleanly urinate, a pierced one must hold his fingers over the newly-made holes or he will trash his friends' bathrooms.

Another approach is to kneel on one leg while urinating. This will allow the opportunity to practice covering the second hole. This avoids the spraying on the outside of the bowl or the floor or wall, or on one's leg.

Case History:

Mark and Henry have been lovers for 2 years. They are both 20 years old. They felt that they had tried everything! They wanted to modify their bodies. They began with tattoos but were not satisfied until they could feel a mutual deeper pain and *pleasure* through piercing. Mark insisted that he and Henry begin with the nipples. They found incredible pleasure when the piercing went through the sensitive flesh. Then they proceeded to pierce their eyebrows, noses, tongues and finally their penises.

Henry became almost crazy when he developed an infection on his nose from the piercing. It didn't stop him. They confessed that the excruciating pain felt when having their tongues pierced was worth it, for the wonderful pleasure it gave them when the stud was rubbed against the penis.

Piercing through the urethra can be especially titillating when urinating.

Anyone reading this must understand that this is an unusual situation. Consider very carefully, before actually having this done.

The ancient mystics of India would lie on a bed of nails. Some have their tongues, cheeks, and/or noses pierced for religious reasons. Some for display or entertainment. They enter into a meditation which enables them to control both pain and bloodshed. It is a rare, mystical experience, where they go off into a limbo state and feel no pain.

American Indians claim to step out of their bodies while experiencing body piercing. He would sit through the piercing and *leave his body* while the needle went through all areas of flesh. This was an indication of great mental and physical strength, endurance and concentration. There was no sexual feeling. The purpose was to prove that he could do it.

In Egypt, slaves were pierced to indicate who they belonged to. Piercing is a means of punishment. Prisoners were hung by their hands. The palms were pierced with nails. Christ was punished in this manner.

Dry Orgasm

In India, the male becomes erect, and is kept so for a month or two. His penis is horribly gorged and swollen. His penis could be 12 inches long and 3 inches in diameter. When the swelling has subsided—his penis is *permanently* twice as large as before. He takes a wooden block to massage it but cannot orgasm with it—it pinches too tightly. If he does orgasm—it is dry.

In Central and South America, weights are placed on the penis of young boys and kept there for months. It becomes lengthened and then gets numb and loses its ability to erect. Why? It makes them highly sexual—but unable to orgasm. This prolongs ecstasy. Orgasm is not the goal. Prolonged sensation is.

Coprophilia/Undinism

Coprophilia is defined as an abnormal attraction to fecal matter. The author, Marquis de Sade, was a proponent of this extreme passion to wallow in feces. No one knows, for sure, if he actively engaged in these activities or fantasized them. He was incarcerated for much of his life for murdering prostitutes. He would have his way with them as he was wealthy and a scoundrel. He writes about tying down a courtesan and forcing her to eat feces. He enjoyed humiliating them while masturbating. He was both sadistic and masochistic.

His many demands placed him into a submissive position. He would send her out to find dog feces and make her push it into his anus while she called him names. "The dirtier the sex, the more pleasurable it is bound to be," Sade quotes. He often preferred to eat the feces of old, ugly, deformed people. The total degradation was his desire. He felt that pleasure requires neither exchange, giving, reciprocating; no gratuitous generosity. It was a completely selfish act. Was he a madman or a genius? They chose to kill him, of course.

Coprophilia is an aberration which has been seen in many of my patients.

Case History:

Martha, 30 yrs., a wife and mother of 3, was unhappy with her husband and sought counseling. She confessed she had never experienced orgasm except through masturbation. After her husband had had his orgasm, she waited until he fell asleep. Only in her fantasy could she masturbate to orgasm.

She imagined her husband would strap her to a stool. He would squat over her and slowly let his feces go into her mouth. He had to tell her how disgusting she was. "Toilet," he would call her. He would make her eat every bit—then command her to clean his rectum. He would proceed to make a sign which said, "Men's Room"; attach a string to it and place it around her neck.

He would then invite anyone of his friends in to use her as a toilet. This caused her to have an orgasm. She felt such guilt at having to think this in order to obtain an orgasm. In reality, this would be considered a perversion, but in fantasy, it is harmless, unless it affects her life. If she could not function as a wife and mother because this fantasy troubled her conscious mind, then it would be deleterious. Since it did not, I told her not to worry.

Undinism

Undinism is defined as an abnormal attraction to urine. This bizarre fetish is far more common than Coprophilia. Often called a "Golden Shower," it is supposed to cleanse the soul of both the giver and the taker. It is a gift to the taker and is well-cherished.

Like Coprophilia, the receiver is usually masochistic and enjoys the humiliation of being urinated on all parts of their body. Many undinists also enjoy welcoming it as a drink. It usually demands that the giver be a *special* someone. Restoring her (or his) individual offering to her lover is a tribute and an honor.

Case History:

Carla was a dominatrix. Her many customers stood in line for a taste of her golden liquid. She said she would drink large quantities of water when she knew she would be busy. Some customers wanted it straight from the source—others wanted it in containers. Some wanted it poured over their heads or on their privates. And, of course, many wanted the supreme humiliation of drinking it.

Case History:

Tom was a homosexual and an Undinist. His form of humiliation needed to be in a public place. He would

go to the Mens room of a toilet and wrap his body
around the toilet bowl. Looking up at his donator and
accepting all the urine which leaked over the sides.
This aroused him to no end. No other form of sex was
acceptable.

Fetishes

An object or person supposed to have magical
powers. Often an object of unreasonably excessive
attention or reverence. The displacement of sexual
arousal of gratification to a fetish.

Apparently Coprophilia, Undinism, Necrophilia,
etc., can overlap into Fetishism, such as the anchor
women described earlier. Her love of the corpse and
the smell of blood aroused her.

Fetishes often combine with odors, tastes, sensation, sounds, and so on. The movie "Blue Velvet" dealt
with a man who was overcome with his sight, feel,
taste of the material—blue velvet. This touching, tast-
ing, seeing this fabric brought him extraordinary sexual
pleasure.

Many fetishes are with fabrics such as hosiery,
panties, brassieres, shoes, etc.

Ladies shoes and boots are a big "turn-on" for a
large number of men. I say men, since I believe that
most very strong fetishes are with men. My practice
has shown me that women seldom have fetishes. I
wonder why?

Case History:

Eric came to me at the age of 22. He confessed that he had always become very excited by women's stockings or shoes. Just to think of them gave him an erection. Normal sex was a bore. But a woman's stockinged leg made him insane with desire to masturbate. He had no interest in intercourse or oral sex—only in masturbation.

I did not wear stockings or high heels during his session. He would be too distracted and we would not continue the session. As long as I wore flat shoes and no stockings, he could concentrate.

When under hypnosis, he related many childhood incidents which most likely led to his fetish. He recalled as an infant, when his mother would place him on the floor to crawl, she always wore nylons and high heels. He remembered crawling up to her feet and rubbing his *diaper* on her ankle. He said he would masturbate. Either his mother was so busy, she didn't notice, or she permitted this. Sounds very strange. He seems to feel that she permitted this as he was an only child and she allowed him *anything*.

Whenever she needed to sort through her nylons to see if they were to be disposed of, he wanted to watch with great enthusiasm. When she placed her hand through the stockings to see if it had a run, he became erect. He was six years old at the time.

Fetishes can become very strong aphrodisiacs. The object (stocking, panty) can hold so much power (significance) as to control the fetishist for hours. He becomes transfixed on the object or person, never removing his eyes or adoration.

Infantilism

A dominatrix related an unusual experience with infantilism. Infantilism is described as an abnormal need for an adult to be treated like an infant. The pampering and desire to be totally passive brings this fetish to life.

Dominatrix Fay, told me about a man who came to her wanting to be diapered. He wanted her to feed him baby food and a bottle of milk with a nipple (of course). He wanted to urinate in his diaper and be powdered and changed. He would make baby sounds like, "Goo-goo, da-da, ma-ma," and so forth. He would gurgle and cry. She had large breasts—he would nurse on them with great delight and, of course, masturbate. She even bought large size baby clothes (purchased at a costume shop) and pretended to show him off to other mothers. If she brought in another dominatrix to play the part of the other lady, that lady would be paid very well. He wanted to hear his (play mommy) boast about her "good baby."

This dominatrix also related other stories to me. One customer who paid her $1,000 to put him into an oven and put the heat on as high as it would go. He was a small thin man who fitted into the oven. He lived through this and emerged with a body full of blisters.

This type of person is an extreme masochist. He told her his mother put him in the oven when he was an infant. Incredible—right? He wanted to relive this incident as it helped him to understand why his mother wanted him to feel such pain. This was a sexual reliving of an incident which always aroused him. This was how his mother showed him attention. The only way he ever felt loved.

Another customer offered her $100,000 to give him sleeping pills and then push him into the river and laugh at him. He wanted total annihilation. She refused to do this at any price.

Many customers begged her to castrate them while in a state of "frenzy"–"fever." This state is described as a "worshipping state" in which the man never took his eyes off of her and would do anything for her. He would repeat again and again, "You are my mistress; I want to give you my manhood." She would prepare the table, knives, bandages, antiseptic. He watched her with an erection. He was completely hers. She was magical; like a saint or god. She acted out a scene of castration, pretending to cut and bandage. This satisfied him. She had to be a good actress and understand he really didn't want to be castrated—he was in an alpha state of extreme arousal.

Then there were the slave men who wanted to be her maid. They paid her to clean her house. Doing the laundry, washing the windows and all menial jobs in the house were carried out by this "House Slave." She was paid highly for this.

Why doesn't he play this game with his wife? Fear of rejection. Fear that she will disrespect him. After all, these men were very successful doctors, lawyers, judges, actors, musicians, priests, rabbis, realtors, business men, etc. All *very* wealthy as she charged high sums for her services. In their profession, they were respected, honored. They carried a great deal of responsibility. That's the word—RESPONSIBILITY. With the dominant, they could put everything into her hands. One lawyer said, "Treat me like the scum I really am—the rotten bastard no one really knows."

Case History:

Jerry was crazy over Paula. Her long black hair dazzled him. He would stare at it, touch it, kiss it, smell it, and so on, for hours, never removing his glance. They were married for 14 years. His foreplay always began with his taking out the pins in her hair, unrolling it, smelling it, stroking it, kissing it, until he had intercourse with it in his mouth.

Paula had her hair bobbed—very short. He could no longer get an erection—no matter what she did. She bought a long wig—it didn't work. He liked the smell of her scalp. They were divorced. Four years later I was surprised to hear Paula say, "We're back together as happy as can be." I said, "I guess your hair grew back." Of course, this was true. No amount of therapy could have changed his desires.

The fetish for smells is primitive. Animals determine all their matings through smells. Estrus in the females causes the male animal to follow her for days until she will accept him. Genital odors are often very exciting.

Case History:

Roger had a fetish for soiled panties and when he couldn't get his girlfriends to offer up their underwear, he spent thousands of dollars on prostitutes who would satisfy his need for their soiled panties.

Unfortunately, fetishes often become so strong, the fetishist cannot have normal sexual relations. His only satisfaction is to indulge in fantasy and masturbate.

Case History:

Allen remembered when he would watch his sister disrobe and wait until he found the opportunity to steal her panties. He was nine when he first took her panties—she was 14. The smell intoxicated him, and led to masturbating on the underwear.

Leather is a very popular fetish. When one wears it, it gives them a look of strength. The masochist who worships leather will enjoy the look, smell and feel, culminating to orgasm. Frequently, this fetishist gives the wearer a certain dominance and power.

Case History:

Ben would pay his mistresses hundreds of dollars to dress in leather. They had to be engulfed in it; leather pants, shoes, hat, brassiere, jacket, etc. In the summer, he had a hard time finding anyone who would accommodate him. His need for so much leather covering, for several hours, was a terrible discomfort, even if she was paid. He needed so much time to get into the feel, taste, smell—thus she almost always gave up. Even air-conditioning cannot help when she had to be in this *very* tight, uncomfortable attire for so long.

The Case of Jeffrey Dahmer

They found his apartment crammed with skeletons, eleven skulls, packages of genitals, preserved and frozen hearts, muscles and innards from his supposed 17 slaughtered victims. His victims were, supposedly, dark-skinned male homosexuals. Had he been black with white victims, the bloodbath would never have gone undetected. This quiet, white man from Milwaukee managed to continue his monstrous activities unnoticed.

Did he have a mental disease, and if so, could he be treated and learn to control himself? Dahmer needed to deny that sadism or hatred of homosexuals and blacks motivated him to murder, dismember and cannibalize his victims. Perhaps he really loved them so much, he wanted to become a part of them by eating them. Maybe his victims were easier to be tempted as they were poor and needed the money he offered them to get them to his apartment.

His need for a zombie-like slave who would always be there for him may have been his reason. Many sexual deviates who molest children (pedophilia) or corpses try to find this escape as they feel inferior. They often have very small penises and are afraid a real live adult person will humiliate or reject them. He claimed it was not a racial, homosexual, hate thing. Well, what was it?

He drugged his victims in order to have a fully cooperative partner. He had no feelings for the human being. Dismembering was a disposal problem. The cannibalism was never explained when he pleaded insanity. His father was a Zombie-like person who never responded to his questions. He was completely unemotional.

Could he be fulfilling incestuous feelings for his father? Was he getting revenge? We'll never know.

Pedophilia

What happens to the child who is sexually molested by her father, mother, priest, rabbi, uncle, aunt, teacher, etc. The victim often feels it is their fault. Until therapy is administered, the young people feel ashamed and dirty. They often lose their religion when it is a priest or rabbi.

Susan Sanoval of New Mexico was molested by Father Kusch. He was a very respected and loved member of society. He took her virginity when she was 16—he was 49. He approached her when she was asleep in the rectory saying it was God's Will and secrecy was expected. She admired him in her innocence. He made her pregnant and made her get an abortion. He also gave her a venereal disease. She continued to have intercourse as he pressured her into believing that it was OK and God wished it. She never told anyone. Too afraid to tell her parents as the priest warned her not to. Today she is suing him and she has become suspicious of all men.

Several years ago in Canada an orphanage was closed when all the children banned together to accuse the priests of pedophilia. Too frightened to tell anyone—who would they tell—they were orphans. They continued to serve the priests as a duty. Victimized, used and terribly alone—their banning together brought the priests to justice and closed the orphanage.

Case History:

Bonnie was an orphan—actually, her mother placed her in an orphanage when her father died. When Bonnie was reading "The Decameron of Boccaccio" (a classic), Sister Florence accused her of reading pornography and placed her into the trial room. The trial room consisted of nuns who sent her to Sister Mary Francis. Sister Francis fondled her and performed cunnilingus on her. Bonnie was told to keep this silent or *she* would be punished severely for *lying!* Bonnie, too afraid of the consequences, obeyed the nun for four years, until she could run away from the orphanage and find a therapist to help her expose the nun and have her fired. The scars are with poor Bonnie— she cannot trust anyone.

Fisting

Fisting is a sexual activity sometimes practiced by homosexuals. It involves gradually relaxing the anal muscles until several fingers and eventually the entire hand (sometimes the forearm) can enter the rectum & colon. The awareness, gentleness and trust recommended for all anal exploration are doubly important. To explore Fisting (giving or receiving), some heterosexual people find this activity very exciting.

Relaxation is most important to accommodate something as large as a hand. There are many anal enthusiasts of Fisting. Homosexuals speak of a "top" and a "bottom." Usually the "bottom" finds it deeply satisfying. The receivers find the sensation of fullness and pressure an ultimate experience. It can be a form of meditation as deep relaxation is necessary. Focused concentration can take several hours. Many rely on large quantities of drugs to relax the sphincter. Physical damage often results as the receiver is not aware of how much he can take when drugged. It is frequently employed with Sado-Masochism to enhance the fantasy of being "taken."

One investigator (Lowry 1981) called it "erachioproctic eroticism." It appears that 40% of those who initiate this practice participate once a week, while 18% were receivers and 37% were inserters. Forty-five (45%) experienced both roles and 24% had actually taken *both* hands at one time. Many men had bowel perforations and needed hospitalization. Occasionally, they reported that it *improved* their general anal health by teaching them relaxation. A rubber glove should be used.

Since AIDS has come into the scene, there is much concern because Fisting causes abrasions in rectal tissues, which provides an easy entry into the blood. If surgical gloves are used, there is less danger.

The recipient should prepare himself by meditation, then lubricate his anus. A dildo is best. Be certain
to press the dildo to an angle so that the object enters
the rectum smoothly. Some people prefer a carrot,
cucumber or zucchini. Wash them thoroughly. It should
be about the diameter of two fingers, and no longer
than eight inches. If it does not go easily, remove object
and start again—relaxing and lubricating. After pass-
ing the sphincter it will reach the pubo-rectal sling
muscle. It stops here, pull it back and move it slightly
at a different angle. Try to find the most comfortable
angle for you. Only two *adults* should practice this
exercise.

Case History:

Ted and George were lovers who practiced Fisting
in their S & M sex. When they began experimenting,
Ted, being the "bottom," felt the urge to have a bowel
movement. They decided to place a plastic shower
curtain on the floor should he need to defecate. Ted
would breathe deeply and slowly. He found that even
though he *thought* he would defecate—he did not. As
he relaxed more and more, it became more pleasurable.
George became very aroused and pushed the dildo so
far into Ted's rectum, he lost his grip on it. This was
serious enough to see a proctologist. They were so
embarrassed. The doctor laughed and said, "Come with
me." He led them to a huge closet which housed hundreds of objects he had removed from rectums. Broom
handles, dildoes, candles, cucumbers, a saxophone, etc.

Why did this doctor keep all these items? Was it to show his patients so that they could relax upon seeing that others have had the same experience?

The doctor talked freely while he was removing the dildo. He told jokes so as to relax Ted. George laughed so hard when the doctor told him one patient asked him to remove a vibrating dildo. The object was so deeply inside him, the doctor said he would need an operation in two days. The patient was so relaxed he asked the doctor if he would please reach up and change the battery. While he waited a few days for the operation, he wanted to enjoy it.

Rape

Rape is the crime of forcing another person into sexual intercourse. Most rapists confess that sex is not the object. The need for *a feeling of power* is commonly the reason for rape.

Case History:

Ralph sexually assaulted 8 year old Nell—he strangled her and discarded her body behind a dumpster. He came to me in deep pain—even tears. He had to tell *someone!*

I am committed to never tell anyone. I would lose my license.

After his second session, he admitted he had to go to the police—he could no longer live with the guilt.

He was given 10 years with no chance of parole and no bail.

He talked about how innocent she was. He described her blonde hair and soft skin and voice. He couldn't stop once he took her behind the dumpster—he *had* to rape her. He felt so *powerful!*

Sex offenders nearly always attack children, women and old ladies. They are weaker. The rapist feels his power for the first time.

Case History:

Ralph lived in rural America. His father was a farmer, his mother had 11 children. Neither had time for him—*ever!* He was a loner and felt that no one cared since he had such a small penis. Raping gave him the feeling he was *big!* Children are small, they cannot distinguish between big and small penises. They cannot fight back.

Sex offenders can easily enter any school unnoticed, take their victim to an isolated place and escape undetected. They will kill their victim not to be punished. Those who are caught will generally receive lenient sentences and spring prematurely due to overcrowded prisons. They are out to commit more crimes. They are very clever and will shop for the best lawyers. Earl Shriner raped boys and girls, then slithered through the loopholes like a serpent who never felt remorse. Some sex offenders can be treated. One national study has indicated that Cognitive-behavioral treatment can reduce audiovision rates—but only by

20%. The offenders remain in prison until advances in treatment are available.

To increase the sentences for sex offenders, the whole penal code would have to be rewritten, with an increase in the sentences for murderers, kidnappers and other big-time criminals. None of the politicians want to deal with this problem. Since prosecutors and parole officials are allowed to slide off the hook—our streets are unsafe. We are all easy prey.

CHAPTER

FOURTEEN

14

TRANSSEXUALS
& BISEXUALITY

A TRANSSEXUAL IS A PERSON who was one sex and becomes another. This is accomplished through the use of hormones, therapy, surgery and sometimes hair removal. Most of these people have a persistent discomfort about their assigned gender. It afflicts one in 30,000 people.

A transgender is one who has finally been able to acquire a body that conforms to their true self. Their goal is to pass successfully in their new identity.

A prospective person with gender dysphoria usually begins with two years of therapy to determine how serious is their desire to be another sex. After all, you cannot change your mind once it's done. A Male to female patient is given estrogen (female hormone). It is injected to make the breasts enlarge and to give the hips a female curve. Finally, the operation is done. This involves cutting the shaft of the penis, removing the erectile tissue, taking the penile skin and pushing it into a hole to form the vagina. The nerve endings are kept intact to give sensitivity to the new vagina. The testicles are already shrunken from the hormone injections. This is removed entirely. A very real looking vulva is formed. A gynecologist would examine her and be fooled. Of course she will always have a male psyche. Except for therapy, the *male* is still aggressive. The testosterone cannot be removed. I have a video of this operation—Amazing! Many transsexuals train themselves to speak in the upper register. Gender is a performance which must be continuously repeated.

Pagan societies throughout history have been obsessed with creatures—from the Sphinx to the Centaur, to the hermaphrodite to the elephant-headed figure known in Hindi as Ganesh—who embodied biological impossibility—their magical, quasidivine status. Freud used the German word *unheimlich*, or uncanny, to describe the primitive religious sensations that the enigmas of existence provoke. To come face-to-face with a transsexual is to encounter the *unheimlich*. They are our Sphinxes, the riddle of our cultural contradictions brought to life.

Berdache groups exist today in many societies. In India, the *hijras* or intersexed men, some of whom are born hermaphrodites, others of whom have their genitals severed (and then bury the organs under a sacred tree), dress in saris, perform rituals for women in childbirth, and occasionally act as prostitutes. Similarly, the *xaniths* of Oman are a socially acceptable group of biological males who wear feminine accessories, perform housework, and also serve as prostitutes.

In the Philippines, boys who cross-dress are accepted as *bayot* as long as their parents identify them as such before puberty.

Collier Cole, head of gender identity clinics in Galveston, claims "there is not one documented case in which psychotherapy has been successful in treating gender dysphoria. It resists such treatment because its origins are genetic or biochemical."

Men and women are distinguished not only by anatomical and chromosomal difference, but also by variations in structure and biochemistry of their brains.

Case History: (Male to Female)

Dora had had 2 years of therapy and estrogen injections, but had not had sex-reassignment surgery. She was gorgeous, tall and slender with beautiful skin and hair. She (he) had a great sense of humor. She was a show girl who drew all the attention away from the real women. She took a vacation in Florida and met Tony, who fell in love with her. She would not have sex with him for fear of losing him. Her reason for not completing the sex gender operation was that she knew many transsexuals who could not have an orgasm when they had the genitals changed.

When three months had passed and he asked her to marry him, he suggested they have sex. She could no longer refuse—she might lose her only love. When Tony discovered she was a man, he was so enraged he literally scalped her. She now has to wear a wig. When she spoke to me and talked about the pain both physically and emotionally, I sympathized with her. When she said, "Well, I don't know why he couldn't overlook that," I roared. It was funny to me and she admitted what she said was funny. How could this young male overlook the fact that she was a he. He wanted children.

Case History:

Sam, like so many transsexuals I've met, was intelligent, lucid, well-read and self aware. He had been a woman too long. He said there's no point in studying the ghetto without studying the conditions which led to the ghetto. He was a teacher and an author. As a girl, his father always referred to her as Sam. The parents had had seven daughters and another daughter dismayed them. She was dressed in male attire. Hair short. Dad took her fishing, gunning and to baseball games. When she got her period, her mother refused to listen to her—"Impossible," she said.

Family members called her Sam. She was sent to an all male college. She could no longer contain herself when she was never attracted to a male. She fell in love with a woman. When they first met they would go dancing. Her lady friend commented on her perpetual erection. Finally Sam admitted to being female and wearing a dildo. Her friend was aghast—ran away never to be seen again.

There were many such incidents like this one. She could not merely be a lesbian—she went for therapy— then testosterone injections and finally the operation. For a woman to man operation—it is much more complicated. The stomach skin is rolled and rolled to form a penis, which is attached to the lower body. The opening of the vagina is closed and the clitoris is extended to the end of the penis. In the case of Sam, the penis became atrophied and fell off four times, and the rolling and attaching had to be repeated. Fortunately, the clitoris remained intact.

Today Sam is happily married. The penis remains erect through a bone placed into the shaft, which remains erect and folds over to her stomach when not in use. She can urinate like a man but can never sire a child.

Case History:

April was a beautiful transsexual. She met Mildred at a lesbian dance. She confessed that she had been a man at one time. Mildred asked if she could have an orgasm. In the case of April—she could *not*. Sometimes this happens due to poor surgery or the mental hangup. Mildred said, "Isn't it important to have an orgasm." April said, "I had to decide which was more important—to be female or have an orgasm—I chose being female." April was now a lesbian. I asked her which sex had the biggest orgasm. She said females. "They're longer and stronger."

Bisexuals

A bisexual is a person who enjoys sex with both male and females. Many indicators will state that you are either heterosexual or homosexual. They say that a bisexual is a homosexual who is closeted (refusing to admit homosexuality). I cannot argue with this, as my experience forever tells me there are many bisexuals.

Case History:

Tall, black young beauty who could have any man she wants. She wants them but needs a woman too. "Where I've been going with Joey, I start dreaming and fantasize about Claire. When I've been seeing Claire—I dream about Joey. I can orgasm with either. It's chemistry, that's all."

Case History:

Sheri loves her husband. He gives her so much sexual satisfaction. She has to see Virginia every week as she loves her body and what she does to *her* body.

Virginia arranged a group sexual party. Three men and 4 women. Sheri experienced them all and enjoyed them all. This was two years ago. Due to the fear of AIDS, she has only her husband and one female lover she trusts. The husband understands and often joins in.

Well, how is that for proof of Bisexuality! There are *many* more such cases.

Homosexuals

Homosexuals are people who enjoy having sex with people of the same sex. The percentage of Gay men in

America is 2.8. Gay women 1.4. Many chapters in this book refer to homosexuals.

There are more Gay people in San Francisco. There are more Gay couples in New York. Many have been forced to become monogamous due to the epidemic of AIDS. Many male homosexuals are promiscuous, needing the approbation of the crowd. Their history speaks of too much mother love and not enough father love.

There is much controversy about whether the homosexual is born this way or whether it is learned. Research has indicated that there is a gene which makes a person homosexual. Narcissism is apparent in some homosexuals. Not only are they attracted to the person who is their own sex, but also to the physical similarity.

Case History:

When Walter and Ed arrived in my office, I thought they were twins. Same height, weight, age, hair and eye color. They had been lovers for 3 years. Walter admitted he loved his looks. He would stand in front of a mirror and praise himself for hours. "How handsome you are. I love your hair. Your eyes are so very, very blue. Um m m—and your penis is perfect!"

Ed was equally enticed by Walter finding him as brilliant as he was. The similarity drew them together as they felt safe being with themselves. There was much trust in this relationship. Of course, they were both lawyers. This is a form of narcissism.

Case History:

Carmen watched Pam as she danced across the floor. She had to dance with her. She saw a halo over her head. Well, Carmen had always seen herself as an angel. Here was Pam looking like an angel. Her long dark hair, shone in the light of the disco. Carmen also had long black shiny hair. Pam's dance style was just like Carmen's. When Carmen danced with Pam, she discovered they were the same age, race, height, and both were nurses. There was instant chemistry. A mirror-image love.

Of course, this is not always true. Many homosexuals are quite different in age, height, profession, etc.

CHAPTER

FIFTEEN

GREATER AGILITY EXERCISES

FOR SUPERIOR SEXUAL PERFORMANCE

As MENTIONED IN FORMER chapters, vitamins can be an aid to better health. If one has little or no energy, certain vitamins such as: B Pollen, Brewers Yeast, Iron, Desiccated Liver, Multivitamin and Mineral, GTF or Chromium, Potassium, Selenium Vita A, Zinc, Octocosanol, Protein Supplement, Vit. B Complex, B12, Vit C, Vit B, DMG, Energen C, Pep Formula, Gerovital H-3, L-Aspartic Acid, L. Citroline, L-phenylanine, Magnesium, Calcium Celate, Asporotate, Royal Jelly, Pantothenic Acid, and Thiamin. Of course, you do not take all of these, consult with your nutritionist.

A proper diet is important. Emphasis should be on fresh fruits and vegetables, grains, seeds and nuts. Avoid energy robbers like sugar, alcohol, fats, caffeine, white flour and highly processed foods. Eat less red meat.

Skeletal muscles usually produce their own effects on parts at some distance from the muscle mass. Thus the power shown in the foot when one runs or rises on the toes is located in the calf of the leg. It is transmitted to the foot through the interposition of a connecting band (tendon) which attaches to bones. If the strong movement in the ankle had to be accomplished by muscle masses located there, the ankle would be many times its size and the presence of muscles there would limit movement.

· Body line is determined in part by skeletal formation, in part by far underneath the skin, but very largely by the size and contour of muscles. The curves of the arms reflect underlying muscles, the lateral swelling of the calves, the margin of the trunk and the cresentric roll of the neck. A person's posture is an expression of muscular action. Although posture patterns are influenced by various forces that play upon man, weakness of trunk musculature often produces most degenerate and disgusting physical types.

Muscles attach to bones by means of strong fibrous bands called tendons. A common tendon is the large central area of the abdominal wall of several muscles. At the lower margin of this tendon are two openings which transmit the spermatic cord vessels, and nerves and are the sites at which hernia occurs. The quadrilateral area of the lower back receive muscle fibers from surrounding muscles.

A "pulled" muscle is a condition produced usually in athletics. The injury is at the junction of the muscle fibers with the tendons.

The second variety of muscle, cardiac, forms the wall of the heart! The left ventricle is thicker than the right. There is a reason for this. The left ventricle pumps blood over the entire body, except the lungs, which are supplied by the action of the right ventricle. The work of a muscle increases its size and strengthens it. The heart is a muscle, and the greater work done by the left side produces a larger muscle than the right. The heart muscle is kept strong by physical activity in precisely the same way that other muscles are.

KEEPING FIT FOR SEXUAL STAMINA

*These exercises must **NEVER** be done without the help of a trained professional.*

Exercise Program: Warm up muscles then stretch muscles to keep fit and limber: By staying in shape it will improve your sexual stamina.

Stretching the groin and hips.

Quadriceps Stretch

Side leg lift for buttock. (gluteus maximus) Kneel down on the floor on all fours. Lift leg at a 90% angle to floor. Lift 15 times and change sides.

Quadriceps Stretch

Back leg Extensions for hamstrings. Extend the leg straight behind you with toes pointed and lift it as high as you can 10 times, releasing it down slightly between lifts. Be sure your head is lifted and your back arched. Besides working the back of your hips, this exercise will also strengthen your back and make it more flexible.

Inner Thigh Stretch
Lay down on your side resting on your elbow for support. Extend your leg up as high as you can and gently press on your calf.

Inner Thigh Stretch
Sit on the floor and raise one leg while holding on to the arch of your foot. Put other hand down for support.

Hamstring Stretch
Grab your ankles, if you can, or your calves and hold for 8 counts.

Sitting Toe Touch

Waist Reaches
Pulling outward and directly to the side, reach your left arm (or right arm) over your head to the right.

Side Stretches
Reach your arm as far upward toward the ceiling as you can for one count. Feel the stretch up your side?

Chest-to-Floor Stretch
Sitting on the floor, open your legs as wide as you can without straining the tendons along your inner thigh. Lean forward towards the floor and get down as low as you can with your hands on the floor in front of you for support.

Praying Stretch
Sit back on your heels. Round down over your knees and extend your arms as far out in front of you as you can. Hold for 10 counts.

Inner Thigh Stretch

Walk your hands back in between the legs, bend your knees into a squatting position and place your hands behind your feet. You'll find your inner knees are resting on your elbows, which should stretch your inner thighs open. Hold for 16 counts.

Waist and Inner Thigh Stretch

Sitting on the floor, open your legs as wide as you can without straining the tendons along your inner thighs. Point your toes. Pull over the right side trying to aim your ear to your leg. Your left arm is pulling directly over to the right while your right arm is curved in front of you. Hold for eight counts. Switch sides.

Knees to Chest Torso Stretch
Hug your knees tightly in to your chest and hold for 10 counts.

Chest to Knee Stretch
Place your hands on each side of your right leg and stretch down, reaching the front of your chest toward your knee. Hold for 8 counts.

Hip and Thigh Release (Releases your inner thigh)
Open your legs wide out to the sides, hold on to your ankles or calves
and hold for 16 counts.

Hip Release
Sit cross-legged on the floor. Round down over your bent legs and hold
for 10 counts, then roll back up.

The Plough

Extend your straight legs out past your head as far as you can with pointed toes and feet touching the floor behind your head. Hold this position a minute or two if you can.

The Plough Side View

CHAPTER

SIXTEEN

16

QUESTIONS & ANSWERS

QUESTIONS AND ANSWERS MOST frequently asked by Patients.

Q: *Why do I lose my erection after I enter my wife's vagina?*

A: This could be due to a subconscious fear of the vagina cured by Psychometry in Chapter VII.

Q: *Why is it so hard for my husband to cry?*

A: Conditioning. Our society frustrates the male's natural desire to cry. This can cause anything from colds to cancer and neurosis to psychosis. (Chapter VII).

Q: *I'm a healthy male, thirty-two years old. Why can't
       I sustain an erection?*

A: It could be diet, starvation of vital vitamins, in-
       ability to express emotions (Chapter I), or fear
       and tension. Inability to express emotions, or
       even the wrong partner (or an uncooperative
       one), overindulgence in alcohol, junk food, to-
       bacco, not enough rest, not enough blood in the
       genital area (explained in Chapter VI), lack of
       stimulation, damage to backbone and fear can
       prevent you from sustaining an erection.

Q: *Why can't I control my ejaculation?*

A: You must masturbate and ejaculate daily after
       Polarity. If sixty years old or ill; two times a week.
       You must use Cranial Cymatics (Chapter VII).
       You must express your emotions in your room
       in Bioenergetics (Chapter VII). You must relax
       with Imagery (Chapter VII).

Q: *I can control my ejaculation when I have intercourse,
       but not orally?*
       I can control my ejaculation orally, but not vag-
       inally?

A: Fear, fear, fear. Need to do Psychometry (Chap-
       ter VIII).
       Fear of teeth, fear that vagina might swallow
       me, etc.
       Fear of the rectum, disease, etc.

Q: *As soon as my girlfriend moves, I can't control my ejaculation.*

A: Practice. Ask her to hold still in between thrusts. Practice Polarity and masturbation (Chapter VII). Practice Imagery (Chapter VII) for relaxation. You cannot have good sex if you are tense.

Q: *My religion states that masturbation is sinful.*

A: Perhaps you can practice Imagery, Polarity, and masturbation in Chapter VII for a month until you have regained control and have a decent erection.

Q: *What is the definition of premature ejaculation and impotence? What percentage of men have them?*

A: *Premature ejaculation:* If you ejaculate before fifteen minutes after you enter. If you ejaculate before your partner orgasms.
    Impotence: The inability to obtain and sustain an erection. As of 1994, in the U.S. alone, of men who have been tested, there are approximately 22 million with impotence and 17 million with premature ejaculation.

Q: *I was sexually abused as a child. Would this have an effect on my being impotent?*

A: Could be. The primary cause of impotence is fear.

Q: *I became impotent after an accident. What can I do?*

A: If you were injured around the brain stem or cervical spin, you may not be a candidate for my therapy, although I have treated some patients whose backbones were fractured and I have almost totally restored their potency.

Q: *I have a full erection when I masturbate, but not with intercourse or oral sex.*

A: Are you subconsciously afraid of the vagina, mouth or rectum? You may be afraid of sex with someone other than yourself. This is performance anxiety. You need Imagery and Psychometry. All in Chapter VII.

Q: *I have full erections in the morning only. Why?*

A: You are still in the alpha or delta state, a relaxed state where you are not thinking of the day's problems.

Q: *I have diabetes and was told by my doctor the medication can cause impotence. It has. What can I do?*

A: I have treated many men with diabetes who were on medication. There has been no conflict and the patient was successful in obtaining an erection. We are conditioned to believe we will be unsuccessful. After all, the penis is only an appendage of the brain. This is called suppressed emotional syndrome (Chapter VI). Ask your MD to reduce your medication, if possible to change it.

Q: *I'm sixty-seven and I'm impotent.*

A: Have you been conditioned to think you should be impotent? There's no reason why anyone should stop having sex when they grow older. You may slow down, but you can have good sex until you die.

Q: *Why do I get good erections with a stranger and not my wife?*

A: Have you ever tried caviar? Your first taste arouses your taste buds. After the thirtieth taste, it's ordinary. Try different scenes. Be Rhett Butler to your Scarlett O'Hara. Be Popeye to your Olive Oyl. Be inventive. Try new things. Have fun with sex. Don't make it a ritual. With my treatment, you will have erections with all.

Q: *I have lost my wife in an accident. I cannot get an erection.*

A: You need time to grieve. Let six months go by before trying again. Do continue to masturbate if you can.

Q: *I have a very stressful job. Could this be causing my impotence?*

A: Yes. Stress will inhibit your sex life. Learn to relax one hour a day. Try Imagery (Chapter VII).

Q: *I ejaculate the moment I enter. What can I do?*

A: Practice Imagery, Masturbation, Polarity, Psychometry, and Bioenergetics (Chapter VII).

Q: *My father, his brother, and my brother all have premature ejaculation. Is this genetic?*

A: No. You have inherited the anxieties. Do you bite your nails, smoke, overindulge in eating or drinking, or procrastinate? Are you a perfectionist? These are all subconscious habits inherited by sons and are a sure sign of a man with premature ejaculation.

Q: *I was always afraid that my mother or sister would catch me masturbating. I have premature ejaculation.*

A: The mind tends to repeat the negative behavior that began at puberty. The first time you had sex with yourself. It continues to do so until you practice Psychometry to eradicate fear (Chapter VII). Rushing the orgasm for fear that a female member will catch you, often causes premature ejaculation.

Q: *I enjoy unusual sex such as anal sex, sadomasochism and homosexuality. Does this cause any problems as far as sexual dysfunctions?*

A: As long as you practice safe sex, why should it cause premature ejaculation or impotence unless fear is involved? Sometimes a patient cannot achieve an erection unless he is in an unusual relationship.

Q: *I have wild morbid sexual fantasies during masturbation or intercourse. Could this harm me?*

A: Why should it? Remember, you can't go to jail for what you're thinking. Sometimes wild fantasies are necessary to obtain an erection.

Q: *I have a small curved penis. Is this why I have premature ejaculation?*

A: You can have any size or shape penis or testicles—even loss of facial hair, enlarged breasts, female body shape—this has no bearing on how soon you ejaculate.

Q: *Because of my small penis I am ashamed to date.*

A: I conduct a group of twenty-five females. When
     they enter a group, they need to fill out an anony-
     mous questionnaire. When asked what they ad-
     mired most in a man, never have they mentioned
     the size of his penis. Men think this is important,
     not women. Most of the female's erotic zones are
     on the outside. Keep that in mind.

Q: *I have an abnormal hormone level. Will hormone
     injections help me get an erection?*

A: I like to treat my patients with vitamins, a good
     diet and herbs.

Q: *I don't have impotence all the time, just sometimes.*

A: This is true of most men.

Q: *Do medications cause impotence?*

A: MD's will say yes. I have had no problem with
     this situation.

Q: *My penis bends in the middle and changes colors on
     erection.*

A: See your urologist.

Q: *I only become erect with a dominant lady.*

A: So, enjoy.

Q: *I'm very shy and have never had intercourse.*

A: Join singles groups. Start feeling comfortable
      around women.

Q: *I lose my erection when I put on a condom.*

A: Very common problem. Are you masturbating
      frequently? You need to activate that muscle.
      You need to sustain an image to keep an erec-
      tion. Try making it a game with your partner.
      Make it part of the foreplay. It can be fun and
      sensual.

Q: *After I have taken your guaranteed treatment, how
      will I know it worked?*

A: When you are with your wife or girlfriend, you
      will know. Remember, it is a permanent cure.
      So, if you're married or don't have a girlfriend,
      the cure will still be good in five, ten or twenty
      years. You only have to know what to do.

Q: *How can what I eat or don't eat cause sexual problems?*

A: Good energy comes from healthy food. Your body cannot function if you load it with junk food (see Chapter VII).

Q: *What are the advantages and disadvantages of masturbating?*

A: The only disadvantage is enjoying it more than being with another person. The advantages are numerous. For premature ejaculation, men store large amounts of semen. An overabundance causes them to ejaculate too soon. When you awake in the morning you need to urinate. Why? Because the bladder has stored urine all night so the urine comes out rapidly and forcibly due to the pressure and the amount. Men should urinate, defecate and masturbate every morning. For impotence: what happens to a muscle when you don't use it? It becomes atrophied and flaccid. The penis needs the activity of the hand muscle since it is a stronger muscle than the vagina or the mouth or the rectum. Use it or lose it. Masturbate daily. Men who are impotent often do not masturbate or masturbate infrequently. Again, masturbate daily. You will stay erect until 104 years old. Men who are forty should masturbate daily. Men who are forty-five to sixty, every other day. Sixty to seventy, every third day.

Men who are seventy and on up, weekly. Masturbation has been known to be therapeutic both mentally and physically. Maybe you are conditioned because of religious upbringing or parental learning to think it is sinful or dangerous to your health. Nothing could be more untrue. You cannot lose strength through masturbation. Your sperm is limitless and you become stronger after rejuvenating it. Religious teaching has been an obstacle in getting men to masturbate. Could God disapprove of such a healthy, natural and pleasant activity? I don't think so. I had a rabbi for a patient who said his religion disapproved of masturbation. He was thirty-two and already beginning to feel impotent. He had been to many doctors who prescribed medication and suggested injections, implants, and so on. What a shame. An implant at thirty-two. He resisted masturbation and became almost totally impotent. It almost broke up his marriage. After resorting to a painful, dangerous implant, he came to me almost suicidal. He felt like a mechanical man and had no feeling in his penis. He finally had the implant removed (which is dangerous, but they do it) and after a few years, returned to me and was ready to follow my program and begin masturbating. I'm happy to say (and so is his wife) that he is having full satisfactory erections. Many patients with hypertension and diabetes have seen a remarkable change in these readings after masturbation. The relaxation of ejaculation lowered their blood pressure and brought down their blood sugar. If guilt plays a primary role after masturbation, please see a therapist. Are you motivated enough to try and lose the guilt associated with masturbation in order to improve your sex life? You need to masturbate if you're having premature ejaculation or impotence, or not sustaining an erection. Masturbation is self-love. Is it so wrong to love yourself?

Q: *Is there such a thing as using my penis too much? Will I grow weaker if I do?*

A: No, you will not run out of semen if you ejaculate frequently. If you get too fatigued, you lose sexual desire. This does not mean you run low on semen. Use your penis often. It will invigorate the rest of your body. Underdoing is worse than overdoing.

Q: *Will a vasectomy affect my sex life?*

A: Definitely not. It is simply a method of birth control. The operation consists of tying the vanus deferens, the tube that carries the sperm from the testicles to the seminal vesicles. Vasectomy interrupts the passage of sperm. The procedure is performed in ten minutes in a doctor's office and under local anesthesia. It rarely requires pain medication. There is no change in penis power,

no change in sensation, no loss in ability to obtain or sustain an erection, and no less pleasure
in the orgasm. Vasectomies have been known to
increase penis power as there is no longer inhibition concerning pregnancy. Partners often feel
a heightened sensation as there is no latex between penis and vagina.

Q: *Will a "cock ring" help me get a better erection or
last longer?*

A: Sometimes men have claimed that the erection
is harder, but as far as lasting longer, no. The ring
constricts the blood outflow, so it is trapped within
the penis shaft and can't escape, thus the penis
has more rigidity. Deep below the skin of the penis
and completely separate from the urinary channel, there are two separate side-by-side chambers
which are cigar-like in shape and length. Each are
similar to a balloon with firm elastic wall. These
chambers or caverns are called corpora cavernosa
and extend from beneath the pubic bone throughout the shaft of the penis and into the head of the
penis. They are filled with very soft spongy tissue. Normally the chamber walls are collapsed
like an empty balloon and the spongy tissue inside it is relatively dry. When blood rushes into
the chambers, as occurs with sexual excitement, it
becomes trapped in the spongy tissue. As the
chambers fill, they expand both in length and
girth. The engorgement results in an erection. The

actual transformation of the penis from a soft flaccid state to a rigid or erect state is a result of blood being trapped in the penile chambers. The change is referred to as vascular transformation. It can be dangerous if blood flow is obstructed unnaturally. Using a "cock ring" encourages blood sludging, clotting or rupture to the delicate sinusoid of the penis. Don't use this. It's not worth the danger involved. Use the squeeze technique described in Chapter VII under Masters and Johnson's technique on masturbation.

Q: *Why am I more virile in the summer?*

A: Cold causes the scrotum to contract and the scrotal sac shrinks. The testicles retract into the inguinal canal. Cold also causes vasoconstriction, a narrowing of blood vessels. Warmth is conducive to super potency. Honeymoons are usually in warm weather. Hot tubs can be dangerous, as the very hot ones may cause the sperm count to lower. The testicles are located outside of the body, because the body is too hot for the sperm to live. Moderation is the key word.

Q: *If I delay my grief, won't I feel bad until I can express it?*

A: Of course, but when you are waiting to express it, hold onto the focus of your feelings as though it was a balloon which you will eventually let loose when you begin Bioenergetics. Keep the feelings light and airy as an appendage you will easily carry until you can let go.

Q: *Does crying really help me in my sex life?*

A: Releasing feelings, expressing anguish, letting go of tears—all of this is healing. When we cry, endorphins are released which help heal your sexual dysfunctions.

Q: *It's hard for me to cry. How can I accomplish this?*

A: Concentrate on a time when you were sad. Concentrate on a sad song or movie. Focus on an incident and let your heart break (see Chapter VII on Bioenergetics).

Q: *When I try real hard to orgasm (I'm female), I find it impossible and give up. What can I do?*

A: Be sure to instruct your partner regarding how you want him to work on you—what position you enjoy most. Ask him for more foreplay. Ask him for cunnilingus. This usually brings one to orgasm before intercourse. If he is hurting you either by his weight or his penis, talk to him.

More foreplay will bring better lubrication. If you terminate your orgasm due to emotional scars, see a therapist. This can be helped.

Q: *I have severe pain when penetrated. I am female. Is this vaginismus?*

A: Probably. It is almost always primarily psychological. Guilt, shame, misinformation about sex, religious taboos, rape, childhood sexual abuse, clumsy unloving partners, painful vaginal examinations, inadequate foreplay, negative attitudes towards the lover, etc., are what cause it. First, see a gynecologist. Sometimes hormone therapy is beneficial. Then, see a psychologist, a therapist or a psychoanalyst. Vaginismus can be treated successfully.

Q: *I have taken your treatment and have learned good ejaculatory control. Will I relapse?*

A: If you follow my instructions, like learning a language or playing the piano, it's yours. No one can take it away.

Q: *What causes a person to fail your treatment?*

A: It almost never happens, but these could be some
     reasons why it might happen:

     1) not following instructions
     2) laziness
     3) fear of success
     4) fear of failure
     5) fear of intimacy
     6) guilt about sexual pleasure
     7) anger toward your partner
     8) fear partner will want marriage since you
        are a good lover
     9) sex is not important
     10) exercise or the excuse "I don't have time"
     11) not taking care of your health
     12) over-drinking, over-eating, over-working,
         drugs, etc.

Once you get into the routine of good living and
following the instructions, most self-destructive
tendencies will disappear.

Q: *Do I need my partner to come to your office? (I'm
     male)*

A: No.

Q: *Do I need her to help with the exercises?*

A: No. Of course, she should be supportive and patient.

Q: *My partner gets "turned off" and seems to want to get it over with. What can I do?*

A: Are you giving her enough of good foreplay? Are your careful about your body being clean and in shape? Are you gentle enough? Do you ask her to tell you what she wants? Are you unfaithful? Have you shaved, brushed your teeth? Would you rather watch TV? Do you embarrass her or make her feel insecure by flirting with others? Do you continually criticize her weight? Do you have the wrong partner?

Q: *I can obtain an erection. How can I sustain it?*

A: You need to focus on an image which excited you, no matter how wild or extreme it seems. Remember, "You can't go to jail for what you're thinking." No one knows what you're thinking, so think it to stay erect. If this doesn't help, see a urologist, then a therapist.

Q: *I'm very shy. I'm afraid to approach women. What can I do?*

A: You're afraid of the big "R"—rejection. Most people are. The only way to overcome this fear is to keep placing yourself around women. Any-where—singles clubs, social gatherings, athletic events, concerts, department stores. Anywhere where women might be. When you approach them, say something like, "You look like an in-teresting person. Can we talk?" If you're in a place where people gather and enjoy music, ask questions about music. If you're at a hardware store, ask her about hardware, etc. Women do go to hardware stores. If and when she snubs you, or insults you, take it as a lesson in com-munication. You may not be the one she's reject-ing. She may be having a problem of her own. Don't take it personally. Try again. Exercise try-ing until maybe the fortieth woman—she may be waiting. Just keep trying.

Q: *Why do I ejaculate so fast!*

A: The reason is tension. Your sexual organs cannot function properly when you are tense and re-leasing adrenaline.

Q: *I've heard about the drug Anafranil for premature ejaculation. What do you think?*

A: I'm against most drugs. They have side effects. Anafranil is a very new drug and no one knows what the long-term effects may be.

Q: *Do I need a partner to help me with the exercises?*

A: Not really. If she wants to help with the Polarity, that would be fine. But, all of this can be accomplished alone.

Q: *How long are the exercises?*

A: You should allow one hour each day to the exercises.

Q: *How long do I need to do them?*

A: Ordinarily, a few weeks are sufficient. Some people leave my office and are prepared to have successful relations immediately. The younger the patient, the better candidate he is for this. Some people wait too long to make a commitment to help themselves. But, it's never too late. When I can help a patient of ninety-four years, I can help just about everybody.

Q: *What are the wrong methods of trying to control my ejaculation?*

A: Drinking, drugs, using two condoms, anesthetic ointments, staying with an unloving partner, surrogate therapy, and especially, not seeking help from a professional therapist.

Q: *When I hold back my ejaculation too long, I lose my erection. What should I do?*

A: When you try too hard to please your partner by holding back, lovemaking loses its sensuous quality. If you become too defensive about your premature ejaculation, you may begin to avoid sex. The chief cause of your problem is tension. You must learn to relax through Imagery (Chapter VII). It is impossible for you to sustain or control your erection or ejaculation when you are under stress. If you become anxious while making love, the body's emergency hormones are released and this reverses the erection, causing the blood vessels to constrict and the blood to drain out of the penis, leaving it soft.

Q: *I can satisfy my partner orally, but when we have intercourse, I come too fast. She doesn't mind. I always feel that she wants me to prematurely ejaculate to get it over with after she orgasms. Does this make me come faster to please her?*

A: Many women prefer oral sex to intercourse (Share Hite's book: *The Hite Report*). This might be the reason for your premature ejaculation. The anxiety felt here is understandable. Talk it over with your partner. You may be imagining it. If not, ask her what you can do to make intercourse more interesting. Be patient.

Q: *I've read about the "Taoist secrets" of cultivating sexual energy. If I masturbate daily, won't I lose this energy?*

A: How ridiculous. The young male stores so much semen, it is necessary to let it go in order to avoid premature ejaculation. When too much is stored, he will prematurely ejaculate. Even the older male should release his sperm frequently for the same reason.

Q: *I'm an Orthodox Jew. The Good Book says kissing my wife's genitals is sinful. She keeps asking me for this and I'd like to please her, but I feel guilty.*

A: If the Good Book says don't *kiss* your wife's
      genitals, why not change the word kiss to *lick*.
      It's a matter of semantics, but if it makes your
      wife happy and you less guilty, what could be
      wrong.

Q: *Since endorphins heal the body, why can't they be*
      injected?

A: Endorphins are *natural* pain killers, a biochemi-
      cal excreted by the brain; an internal morphine
      which is ineffective when produced artificially.

Q: *Since endorphins heal, can they also act as a preven-*
      tative?

A: Research is currently being done on that.

Q: *I have been happily married for twenty-five years.*
      Our sex life has been wonderful. Recently, my hus-
      band cannot get erect. He asks me to do fellatio on
      him. I'm embarrassed, but would like to please him.
      Please help me.

A: Perhaps you have been conditioned to think that
      the penis is not clean. Try to overcome this
      (therapy may be needed). The penis is usually
      as clean as any other part of his body. Love it as

you love any other part of him. If you fear the semen, this may make you uncomfortable, ask him to *withdraw* before he ejaculates or spit it out. Practice makes perfect. It may be fun. Try it.

Q: *I have soreness at the entrance of my vagina. It is so painful when my husband tries to penetrate. Why?*

A: I'm not a gynecologist, but if you had stitches there after childbirth, you may be sensitive and require lubrication before intercourse.

Q: *I have pain only when my partner thrusts deeply during intercourse. Why?*

A: There may be pressure on an ovary which may be caused by a tipped uterus. Try different positions. If the condition continues, see a doctor.

Q: *Why am I so dry? It is painful to have intercourse.*

A: Sometimes menopause brings about a thinning of the vaginal lining and a decrease in natural lubrication. Your doctor may prescribe hormone replacement therapy. Estrogen cream can help, but a lubricant may be all that is needed.

Q: *My vagina seems too tight to penetrate. Why?*

A: This can be due to an allergic reaction to vaginal douches, deodorants, bath additives or spermicide. You may have a mild vaginal infection. You may be tightening your vaginal muscles involuntarily due to fear. Do you have hemorrhoids? Stop some of the additives first, then see your doctor if this persists. You may have a condition called vaginismus. This can be physical or psychological.

Q: *I had many satisfactory sexual encounters before I was married. My husband of twelve years cannot arouse me. I truly love him, but whenever it's time for sex, I do it as a duty. Why?*

A: Your duty is to tell your husband how you feel. Explain what you want him to do to you sexually. It could be the answer to better relations.

Q: *Is masturbation physically or mentally harmful?*

A: No.

Q: *I feel guilty about masturbating. Why?*

A: Most women masturbate. Get in touch with your body and mind. Consult a therapist. Could God want you to give up anything so pleasant?

Q: *I can always climax through masturbation, but never through intercourse. Why?*

A: In Dr. Hite's book, *The Hite Report*, she tells of hundreds of women who need to hold their legs tightly when having intercourse. Try a side position where you can hold your legs close with your partner behind you and try touching or having him stroke your clitoris at the same time.

Q: *I have itching around my vagina. Why?*

A: It could be allergies, diabetes, infection, or many other things. See your doctor.

Q: *I always fake orgasm when I have intercourse, but not with oral sex. Why?*

A: This is a common complaint. Women are often so concerned about pleasing their partners and hurrying their intercourse that they often neglect to inform their partners about their own needs. Communicate your wishes with your partner. When having intercourse, think of nothing else but your climax. Don't think about your weight, your hair, your need to care for the kids, etc. When you're having intercourse think of only sexual things—fantasize—think the wildest things. Who knows what you're thinking! That's where all the feeling is. If your partner neglects you, takes you for granted, is unfaithful, unclean, alcoholic, etc., look for a therapist or leave him. You have a right to pleasure.

Q: *When my husband died I became paralyzed emotionally and couldn't function sexually. Why?*

A: I wonder if the human animal has a self preservation instinct which either paralyzes or causes amnesia in order to give the mourning person a chance to recover. Mentally and physically you may not be prepared to face the shock. Your system puts a halt to everything, to give you time to absorb the hurt. After you have accepted the loss, you may mourn with tears to bring about a catharsis. This is imperative. You must mourn eventually. Read Chapters VI and VIII on *Suppressed Emotional Syndrome* and grief work.

Q: *It's been nineteen years since my son was killed in an accident. Why do I still feel I might have done something to prevent it?*

A: Guilt almost always shows itself after a death, especially that of a child. It's natural to feel the way you do. It's time to move on now. See a therapist. Join a bereavement group. Call the hospitals. They will direct you. You need to mourn and move on. Stop thinking about it. Stay busy. That's important. Talk to friends. Get involved. Cry.

Q: *When my mother died I joined the Buddhist religion.*
      Why did it have such an impact on my life?

A: The Buddhist religion teaches us that all things
      happen for a purpose. Did you learn anything
      after her death? Did it change your life?

Q: *I went to a medium to try to reach my wife. It was*
      surprisingly effective in helping me feel that my wife
      was a happy spirit. Why does this type of service
      help?

A: Who knows? Some clairvoyants can be most
      beneficial in helping people communicate with
      the dead. I have known many mourners who
      have found much peace in feeling that all is well
      with their loved ones who have departed.

Q: *I've developed an ulcer and insomnia since my father*
      died. Does this have any bearing on my lack of ex-
      pressing my grief? I've been conditioned to be un-
      emotional.

A: You must express your feelings. You need to cry,
      scream, let go. (See Chapter VIII) on expressing
      grief. Stop making yourself sick. Express.

Q: *I'm a male who feels foolish crying, even privately.*
      How can I cry?

A: Our culture teaches males to be unfeeling. "Don't
      cry, don't be a sissy, be a man." Does this mean
      you should suppress your tears? Women think
      men are sexy when they cry. If you do not ex-
      press your grief, you're open to all kinds of ill-
      nesses from neurosis to psychosis—from colds
      to cancer. Go ahead and cry. Think of a sad movie
      or a sad song and let yourself cry. See Chapter
      VII on Bioenergetics.

Q: *When my two children died in a house fire, I thought
      I should have also died. I was so angry. How can I
      express my anger at God? I can't speak to anyone
      and can't work. Help me express anger.*

A: In Bioenergetics (Chapter VII) I show you how
      to express your anger. Do the exercises and make
      yourself get angry and let go. Face the fact of
      death. It is not easy, but it is possible through
      Bioenergetics.

Q: *I'm a recent divorcee. Just four months ago, my hus-
      band left me for another woman. I'm full of self pity.
      How can I adjust to a new environment?*

A: It's natural to miss the things you shared with
      your husband. Divorce is a death. You must
      mourn it just as a death. As you search for some-
      one to fulfill your needs, remember you must
      mourn this death first. It takes time. Remember,

"This too shall pass" from the Bible. You are angry and sad. Release through Bioenergetics (Chapter VII).

Q: *I'm a fourteen year old male who just lost my parents in an accident. How can I accept this?*

A: Well, you really can't. It's tough. You're so young. I hope you have friends and relatives around to help you. It's natural to be angry, frightened, helpless, sad, and so on. 'If only they were back,' is what you say. You would not be so demanding, self centered, lazy, smart allecky, etc. You keep blaming yourself. No amount of rationalizing will bring them back. Time will heal the wounds. Maybe not entirely, but they will lessen as you become busy. Busy is the word. Get busy. That's so important. Get busy and stay busy. Deliver newspapers, join a school activity, see a therapist. You need support from your friends and relatives. Reach out. You must mourn. Try to detach yourself from the dead. You need to find hope in daily activities. It hurts right now, but the wounds will heal. Someone will fill the gap that was left by their death. Do not turn your anger inward. It was not your fault. Don't be irrational and distort things. You will survive. You will be in shock and it may all seem incredible. You may feel numb, overwhelmed, anxious, and so on. There are normal reactions to sudden death. Your mouth may feel dry and you may lack energy. Your muscles may tighten. You may be oversensitive. You may feel breathless and depressed. You will be confused and have difficulty sleeping.

These are all things that happen after a death, especially with someone as young as you. They don't all have to happen, but they may. So, just don't feel funny if they do. You may withdraw socially, be absentminded, dream of the deceased, even have nightmares. Read the chapter on Bioenergetics and practice the exercises to help you express anger and grief. Cry in your room. Cry, cry. Tears have a wonderful healing power. Tears have mood altering chemicals. Join a bereavement group of people your age—for teenagers. It will help you heal. But, cry as much as you can. You must relieve your stress. Crying does this. Let your heart break. People will understand. You have had a great loss. So, cry.

Q: *I have a two inch penis with a half inch circumference when erect. I'm horrified to be with a woman. Is there a place to have a penis enlargement?*

A: I do not encourage this. This involves major surgery even though it is performed in a doctor's office and only takes 45 minutes. The dangers outnumber the advantages. The advantage is, of course, a somewhat larger penis and perhaps greater self-confidence. Dangers include the pos-

sibility of a wobbly and distorted penis, nerve
damage, bleeding, infection, and even death. If
you opt for this invasive procedure, call your
local hospital and ask for a plastic surgeon. Talk
it over with at least four plastic surgeons and
demand referrals. There are some men who have
been happy with penis enlargement and many
who have not. Remember, a woman receives her
physical stimulation primarily from the outer 2–
3 inches of the vagina and more importantly,
from the clitoris. Length and width is not important to her.

Q: *When I was 4 years old, I knew I was in the wrong
body. I'm a 28 year old male with the psyche of a
female. I've always admired females and felt I was
one. How safe is a transsexual operation? Where do
I go for this?*

A: Seek a plastic surgeon. Call your local hospitals.
Meet with at least four before you decide. Demand credentials and referrals. I have talked to
many transsexuals. Some are happy, many are
not. There are two years of psychotherapy, hormone injections and then surgery. It seems that
the penile skin is opened and the erectile tissue
is removed. The penile skin is then used as a
vagina. A hole is made in the body and the skin
is pushed into the hole like a glove. The exterior
vulva is arranged to look so much like a real
woman, even your gynecologist can be fooled.

Some transsexuals claim they can no longer achieve an orgasm with the new genitals. Some can. One transsexual told me that he had to choose between having an orgasm and being female. He gave up the orgasm.

Q: *Why can't I have unlimited sex?*

A: You can! If you are age 16–40, wait one hour before resuming intercourse. It takes that long for the sperm to replenish itself. If you are age 41–60, wait two hours. If you are age 61+ wait two days.

Q: *I'm female; I have multiple orgasms; why can't my boyfriend?*

A: He can but not usually immediately. Some younger men can in a few minutes. Depending on age and mood, some men can in a few hours. Many men hold back to experience more pleasure on a big orgasm.

Q: *Is there a G-Spot?*

A: Yes. It is located in the upper inner wall of the vagina, close to the pubic bone and the clitoris (which is on the outside). Women describe it as the "inside" of the clitoris. Pressure on the area

produces a physiological stimulation different
from the clitoral orgasm. Try touching it with a
gloved finger (not to scratch yourself) or dildo.
For intercourse, try the "on top" position so as
to place your finger on the G-spot. A man; gynecologist named Ernst Grafenberg discovered this
"zone of erogenous feeling."

Q: *Can you put a condom on a flaccid penis?*

A: No, he needs a partial erection.

Q: *What is a Dental Dam and how does it work?*

A: A Dental Dam is a condom for women. When a
woman is having oral sex, some feel safer with
one. It is a square of thick latex about six by six
inches sold by dental supply stores or your den-
tist. Some stores sell them in the condom area.
They come in all colors. They look great, but
they don't feel great. It is made of thick material
and it's hard to feel anything. The man has a
really hard time finding the clitoris as the rubber
is so thick.

Q: *Do condoms feel OK?*

A: It's like being naked as compared to clothes. I
don't deny the need for them. Try focusing on the

many sensual pleasures surrounding it. Try having your girlfriend put on the condom with her mouth; it's awesome. Make sure it's the dry kind. If lubricated, the taste can be disconcerting.

Q: *Can a condom keep me (male) from prematurely ejaculating?*

A: Possibly!

Q: *Can a woman climax without any physical contact?*

A: Yes; 64% have been able to think their way to ecstasy.

Q: *Why does it take so long for my wife to orgasm?*

A: Talk to her; ask her what she likes. Be patient and gentle. Act out fantasies; be original. Take time to talk sexy and touch her all over.

Q: *If I had a larger penis, would I be better equipped to please women?*

A: It has been proved over and over again, that penis size does not make her feel sexier. Her feeling is primarily in her clitoris which is outside the vagina. Some women prefer a man with more girth than length. But if you stimulate her clitoris you've got it licked (no pun intended).

Q: *What is sexual chemistry?*

A: An irresistible pull to a member of the opposite (or same) sex. It can be a sight, a sound, a smell, a touch, etc.

Q: *Why does my husband have difficulty maintaining an erection?*

A: Neurological or hormonal impairment can interfere with erections. If circulation is the problem, *"Polarity,"* (an exercise described in detail in Chapter VII) is the *answer.* If it is psychological, treatments in Chapter VII will guarantee sustained erection. Only a doctor can determine this.

Q: *Does a woman feel sexier as she gets older?*

A: Yes, in many cases. As the testosterone increases, her libido shoots up-up-up. Some women experience dryness during menopause. A gynecologist can help here.

Q: *I've heard that fingerprints can determine if a man is homosexual. Is this true?*

A: Researchers at the University of Western Ontario claimed that homosexual men have more ridges in their fingerprints of their left hand, than heterosexual men do. Fingerprints are completely developed in human fetuses by the 16th week and are determined by heredity. This suggests that sexual orientation is somehow determined by pre-natal events.

Q: *Is there a "Gay Gene?"*

A: A National Cancer Institute scientist claimed that homosexuality was inherited from the mothers of gay men. This has not been proven as yet.

Q: *Where are the largest number of gay couples in the country?*

A: NYC #1; San Francisco #2.

Q: *What is Tumescence?*

A: Nerve endings.

Q: *What are some of the reasons men ask women to marry?*

A: Men wish to be accepted for themselves. Most men seek emotional and a clear understanding of the lady. Men enjoy the physical differences. Their mysteriousness, their smell, their softness, their voices, her body (of course). Most men talked about a lack of competition in women as compared to a man friend. "Women are better listeners; they're more interesting." Most of my male friends talk about sports and cars and money. Women have a whole world to talk about as there is not so much competition. Men don't care for nagging, smoking, jealousy, untruths, and women who persist in talking about marriage. Men want to be the aggressors. Sex is viewed as a poor reason to get married. It's important, but not the big reason. Compatibility was the No. 1 reason for men to ask a woman to marry him. He wants from her what he cannot do well on his own.

Q: *I've (male) made the commitment to get married—so why do other women look so good to me?*

A: It's a last minute fling-thing. Suddenly, you're going to be with one woman—what about all those other women? It's natural to feel this way. It probably won't last, so don't worry.

Q: *Is it true that in some countries men offer up their wives to their male friends?*

A: Yes, in some places in Russia (and other places) a man will tell his friend, "Everything I have is yours; my wife is here to serve you." He then disappears until the next day. The wife *must* obey. Marriage is not based on a concept of mutual consent.

Q: *What is a hermaphrodite? What can be done to correct this?*

A: A person born with both male and female genitals. Dr. John Money, of Johns Hopkins University has done many studies on hermaphrodites. When the malformation occurs, the surgeon will determine which genital is more fully formed and remove the other one. For instance, if the vagina is fully formed and the penis and testicles are very small, the doctor will remove the male parts and give the infant estrogen. And vice-versa. If the infant is more fully male, the vagina is closed and testosterone is administered. Some famous actors, actresses, politicians, priests, lawyers, etc., began as hermaphrodites and were turned into the sex which was most dominant at birth.

Q: *My penis is less than one and a half inches when flaccid. When erect, only 3 inches. I'm desperate; what can I do? How safe is the operation? How much does it cost? Does insurance cover the operation?*

A: Yes, you can be helped. I would not advise this for men with penises over 5 inches, but *you* could use some help. Dr. Melvyn Rosenstein has an office in Scarsdale and Los Angeles. To lengthen it and give it girth—the cost is $5,900.

   An incision is made above the penis just below the pubic bone. The ligament is severed. This ligament attaches the penis to the pubic bone. It can gain about 2 inches; no more. It can change the angle of the penis. Instead of pointing upward, it may point downward or horizontally. The scar tissue can reconnect the severed halves of the ligament and pull the penis back into the body. He may end up with *less* length. There are tales of mutilation—even death. But phallomania persists and often the man with a microphallus will risk the possible dangers for the reward of a larger phallus. Dr. Rosenstein has performed over 35,000 penile augmentations. I don't believe insurance covers this, as plastic surgery is elective surgery. He does offer financing. Most patients have been happy and feel more self-esteem; especially in the locker room. They want to have more sex.

In India men tie heavy weights to their penises when young. They stretch them to as long as eighteen inches. They are so long, they need to tie them into a knot in order to get them into their loincloths. American men confess to tying 3 pounds of weights to their penises when they are alone. In a year one man gained 2 inches.

Most women do not complain about penis size—some do. One woman claimed a boyfriend was *so* nice, but his penis was too small to satisfy her. Some women prefer men with wide penises as it strokes the clitoris better. When they are too long, it often hurts where it thrusts against the cervix. Too much of this has been known to cause cervical cancer.

Q: *Why are so many men drawn to Dominatrixes?*

A: Many men who are attracted to Dominant women (who charge for their services) because it is Safe Sex. The role play is safe. There is no intercourse or any sexual contact. It is a game.

Many men feel the need for punishment and humiliation due to incidences relating to their past. Mom was dominant. She demanded a perfect son or daughter to obey her every command. Dad was a rigid disciplinarian who whipped the child repeatedly. Aunt Mary, dressed her little boy in girl clothes and spanked him. The girls in school would push little sissy to the floor and step on him and call him names. His sister would wet the bed; she enjoyed the warm feeling and the smell of urine. Men and women want to reenact the incident with a person who will help them understand themselves and give them much pleasure reliving the incident.

Q: *Why do men go to prostitutes when they have a willing wife?*

A: Vanity. The prostitute will do anything he wants her to do. Often the wife will refuse and think it perverted. Many men need more sex than the wife will provide.

Q: *Some men go in for a massage and often have the masseuse masturbate them. What is the reason when intercourse is available?*

A: It's safe. They are cuddled. They can be passive. Lots of men really enjoy being masturbated.

Q: *Pornography is so boring to me. I'm a female with a strong sex drive.*

A: It can rekindle a dying sex life. It gives one new ideas. It stimulates one further toward desiring intercourse.

Q: *Why is Phone Sex so popular?*

A: It's safe. The caller remains anonymous. He can say anything and not feel embarrassed. Like going to a prostitute—there is no need to wine and dine. It's simply mechanical, convenient and SAFE.

Q: *Why does Miscegenation occur when the people involved know it may cause the offspring to be uncomfortable?*

A: This is not always true. Many couples of different races are attracted by the very color, smell, attitude of their partners. There, we have *chemistry* again.

Case History:

Jenny and Gus were married. They had been going together for 3 years. Jenny was black; Gus was white. Gus confessed that he felt excited by the fact that his family and friends thought he was crazy. Gus was a very successful doctor, handsome, witty— all the things a girl could want. He chose Jenny. He felt sexually *inferior* to her because she was black. It made him feel lower. This relationship brought out his need for humiliation. They have 3 children. The children are well regarded as they are wealthy and educated.

Q: *What about Romance? What has happened to Ro-*
      mance? I'm 70, a male and as much in love with my
      wife as I was when I met her fifty years ago. I still
      bring her flowers and we dance to love songs.

A: Gone are the days when Romance reigned. We
      are becoming a practical people. Children want
      to be different from their parents. They rebel with
      clothes, language, tattoos, piercings, etc. Bring-
      ing a girl flowers, is not *cool.* It's sad. Perhaps
      this fear of AIDS will bring a new generation of
      adults who will cherish poetry and monogamy.

Q: *My partner has an orgasm when I kiss her neck, feet,*
      palms. How do you explain this?

A: Some people are highly sexed. You must be some
      kisser! The skin is the largest organ on your body.
      It is highly sensitive. European men have related
      the interesting fact that kissing a woman's hand
      is meant to be more than a salutation. Often there
      is a licking as well.

Q: *Should I douche every day?*

A: There is a lot of controversy over this question.
      Some doctors feel once a week is fine. Some feel
      it is unnecessary even deleterious as it destroys
      the natural secretions. As long as the genitals are

bathed daily, daily douching does not seem to be recommended.

Q: *My wife needs me to thrust very heavily in order to orgasm. We are often both sore after sex. How can I satisfy her?*

A: Be sure to lubricate with KY Jelly. Go to a sex store and purchase a vibrating dildo. Some dildos are *very* hardy. Why not start out with this and end up with you? Spend more time on foreplay for lubrication.

Q: *My husband is uncircumcised. When I pull down the foreskin to fellatio him, he has an odor and some pubic hair stuck in there. It makes me sick. What can I do? I love him. Otherwise he's* very *clean.*

A: Ask him to please check his penis before he asks you for oral sex.

Q: *My new boyfriend asked me if I was a "swinger." I'm 16 and not experienced. Where is this done?*

A: A "Swinger" is one who goes with a "group." In other words he wants to know if he could bring another man (men) or woman (women) to have sex with both of you. It's done just about

anywhere and everywhere there are willing subjects and a place to party. In large cities, some
newspapers such as *Screw* or magazines like
Single Swingers are available on newsstands.

Today, with the advent of AIDS, this is not
recommended. How can you be sure your boyfriend is safe. If he is a SWINGER, he has probably been with many men and women. Make
him take an AIDS test and be careful!

Q: *My husband is a great lover, but I find that I need
variety! I feel guilty about the extra-marital affairs I
have.*

A: Does your religion forbid this? If so, talk to your
clergy or go to a therapist. If you are feeling
guilt over your actions, it can cause problems.
Many patients have expressed a deeper feeling
for their mate after an affair. For people who are
inexperienced, an affair can be educational. For
those who are insatiable, it can be a relief. Be
sure to use protection.

Q: *Tell me the difference between homosexuality and
bisexuality.*

A: A homosexual is attracted to only people of their
own sex. A bisexual is attracted to both sexes.
They fall in love with a person, not a gender.

These terms were invented in the 1900s.

Q: *I don't want a long-term relationship. I love one-night stands. Am I normal.?*

A: Normal is only a word. Whatever makes you happy is normal. So long as you do not hurt yourself or someone else. Most people who enjoy one-night stands are those who fear intimacy or are commitment-phobic.

Q: *My girlfriend says she thinks vampires are "sexy." Is this sick?*

A: "You can't go to jail for what you're thinking." Engaging in either sucking blood or having it done to you is highly unusual and can be delicious. Remember, there is a malady called AIDS. If you have an abrasion in your mouth while sucking—the blood can lead to danger.

I've heard people say they enjoy the taste of blood. There is a sexual connotation insofar as the *taking* of another life—giving fluids may arouse the sadistic nature of the taker. Having it done is considered masochistic. The giver is giving her life-sustaining energy. I've heard stories of people who sexually enjoy watching wild animals eating blood-giving flesh. Frank Langella portrayed Dracula on stage. The critics called it outrageously sensual as he would aim for the genital and end up at the neck.

Q: *Is incest very common?*

A: Unfortunately, yes. In all countries, races, eco-
      nomic groups, and sexes. Most common is fa-
      ther and daughter.

Q: *I'm 150 lbs. overweight. My husband of 20 years is*
      of normal weight. Recently, I lost 150 lbs. He now
      finds me unattractive. I'm so happy with my new
      figure, but distraught over his lack of interest. What
      can I do?

A: A patient related a similar story to me. After
      working so hard at reducing, her husband di-
      vorced her and remarried a heavier woman.
      Some people feel threatened when a spouse looks
      good. Other men are attracted to her. This is a
      basic insecurity and he needs to see a therapist.

      I had a male patient tell me he could no
      longer get an erection with his thinner wife. He
      felt like he was on top of a skeleton. He loved
      her fat.

      Perhaps it's time to talk to your husband or
      ask *him* to get help.

Q: *Why is it painful to urinate after sex?*

A: If your doctor can find nothing wrong, it could
      be an inflammation of the urethra. This can hap-

pen when fingers or the penis is inserted into the vagina, which could introduce a bacteria in this area. This is called *Urethrites*. Consult a urogynecologist. Try to urinate before and after sex to help keep bacteria from traveling to the bladder. Drink lots of liquids. Make certain you are well-lubricated before intercourse.

Q: *I have a hard time focusing on sex during intercourse. My mind wanders. What can I do?*

A: Tell yourself you will think of things other than sex—tomorrow. Take your time during foreplay. Have a longer foreplay. If you don't get turned on with your partner, find another or—role-play that you are Scarlett O'Hara and he is Rhett Butler. Let your imagination go wild. Think anything that gets you aroused. Even if it's a fantasy that he is not into—it's OK to fantasize.

Q: *I'd like to purchase a vibrator but which type should I buy?*

A: There are various sex shops if you're in New York. "Come Again" (in Manhattan); "Eve's Garden," etc. Ask the proprietor to explain the different types. They are usually cooperative. Don't be embarrassed; they do this every day.

Q: *What are a man's erogenous zones besides his geni-*
      tals?

A: Ears, nipples, neck, stomach, rectum; and don't
      forget to talk dirty. Many men like this.

Q: *Men think of sex all the time. Why?*

A: It feels good, great! It's a stress release; it reaf-
      firms their masculinity; it's an ego boost; they
      want to spread their seed; it's a primitive in-
      stinct and the most personal thing you can do
      with someone.

Q: *I'm ashamed to take my clothes off—my breasts are*
      so small.

A: Many women and men are self-conscious about
      their bodies. We need to have more self-esteem.
      Your partner does not only love you for your
      body. You have a good personality, you are fun,
      smart, interesting. Why not focus on the posi-
      tive—eliminate the negative.

Q: *Are the sex issues facing this generation "mind-bog-gling?"*

A: Every generation has problems.

Today it's AIDS
Yesterday it was DRUGS
Years ago it was PREGNANCY

Q: *It depresses me that I'm so afraid of sex.*

A: You were BORN sexual and *learned* to hide or fear the vital part of yourself. As you reject a part of yourself, you reject your personal expression and vision.

Reasons/Excuses for Not Enjoying Good Sex

I'm too old
I'm too tired
I don't have time
I've got a headache
I'm sick
The last time I did it, I was too quick or couldn't erect
It's too painful
I can't find the right partner
I hate my husband
I hate my wife
I can't orgasm
He turns me off
She turns me off
I'm not drunk enough
He doesn't do it right
She doesn't' do it right
Fear of pregnancy
Fear of disease
Medication stops me from erecting
I'm not turned on
I can't afford therapy
I'm too scared
I'm too angry
It's boring
Sex is dirty

Dispel them all . . . You can be helped!!

List Your Own Excuses:

(a)

_____

(b)

_____

(c)

_____

(d)

_____

(e)

_____

(f)

_____

(g)

_____

(h)

_____

(i)

_____

(j)

_____

(k)

(l)

(m)

(n)

(o)

(p)

(q)

(r)

(s)

(t)

(u)

(v)

(w)

(x)

(y)

(z)

Conclusion

Body/Mind Inseparable

What goes on in the body reflects what is happening in the mind. The exercises I teach you show you how to unlock negative energy and re-own yourself. When you work with body and mind to help you realize your potential for pleasure and joy, you make your vulnerability become your strength. When a depressed person's energy level is raised with vital actions, and thoughts, his depression disappears. Remember, when you're depressed; get busy...do something. You'll forget your problems; at least, for a while. The immediate result is a sense of security. By grounding the patient, he becomes more fully identified with his animal nature which includes his sexuality. My exercises allow the body to move with few limitations. Through special movements and body positions the patient in Bioenergetics therapy gains a deeper contact with his body and a better feeling for it. Sobbing releases the tension in the throat and also opens the belly. Crying is the basic mode of releasing tension as anyone can see in a baby crying when he is frustrated. This creates an insurable tension in a releasing outlet. Aren't we all infants, at heart? Cry, laugh, get in touch with your body. My treatment aims at feelings as emotions erupt spontaneously and huge tears come forth. The goal is to let go, to receive pleasure.

You are shadowing your natural brilliance into layers of conditioning and appropriateness that may keep you from this natural source of satisfaction and improvement. Find a way to unmask yourself and uncover and express the parts of you that have been secreted away from others and possibly from yourself.

Find a good sex therapist who can possibly help you through self-exploration and liberation, and step into the wonder of the unknown.

Search for the right partners; he or she is out there. Don't despair. Someone who can free you from the limits you've placed on yourself and your life. We all want love, acceptance and safety. Have fun, but be careful!

There is much joy when you meet the right person. Explore to find the spiritual bond of trust and closeness. So what if you lose; the journey is worth it. Look at what you've gained. Drop your expectations and receive life as it is. You will grow in competence and access your inner knowing and invite capacity for sex.

Sex Related/Health/ AIDS Services

The AIDS Health Project
Box 0884
Dept. P
San Francisco, CA 94143-0884

AIDS Atlanta
1132 Peachtree Street, NW
Dept. P
Atlanta, GA 30309

AIDS Foundation Houston
3927 Essex Lane
Dept. P
Houston, Texas 77027

American Foundation for AIDS Research
Box AIDS
Dept. P
NYC 10016

Project Inform
347 Delores Street / #301
Dept. P
San Francisco, CA 94110
1-800-822-7422

AIDS Action Committee
661 Boylston Street
Dept. P
Boston, MA 02116

National Association of People With AIDS
2025 1st Street, NW
Dept. P
Washington, D.C. 20006

Where To Go
For Help

Alabama
Birmingham
University of Alabama at Birmingham, Mental Health
   Studies.

Arizona
Tucson
University of Arizona School of Medicine, Sexual Prob-
   lems Evaluation & Treatment Clinic.

California
Long Beach
Center for Marital and Sexual Studies.

Los Angeles
University of California at Los Angeles Medical School,
   Human Sexuality Clinic

University of Southern California School of Medicine,
   Sex Therapy and Marital Counseling Clinic, LAC-
   USC Medical Center.

Orange
University of California Irvine Medical Center, Depart-
   ment of Family Medicine, Sex and Marital. Therapy
   Clinic.

San Diego

University of California San Diego Medical Center, Department of Reproductive Medicine, Gender Dysphoria Team.

San Francisco

University of California San Francisco School of Medicine, Department of Psychiatry, Human Sexuality Program.

Stanford University Medical Center, Department of Psychiatry, Biofeedback Stress Reduction Clinic.

Colorado

Denver

University of Colorado School of Medicine, Human Sexuality Clinic.

Connecticut

Farmington

University of Connecticut Health Center, Department of Psychiatry, Sexual Education and Treatment Service (SETS).

Hartford

Sex Therapy Program

Stamford

Stamford and St. Joseph's Hospitals, Stamford Center of Human Sexuality.

Westport
Hall-Brooke Hospital, Hall-Brook Family Therapy &
    Human Sexuality Program.

District of Columbia
Preterm Center for Reproductive Health, Sex Therapy
    Unit.

Georgia
Atlanta
Emory University School of Medicine, Department of
    Obstetrics/Gynecology, Emory University Clinic.

Illinois
Chicago
Cook County Hospital, Social Evaluation Clinic.

Maywood
Loyola University of Chicago, Sexual Dysfunction
    Clinic.

Springfield
Southern Illinois University School of Medicine, De-
    partment of Obstetrics/Gynecology.

Indiana
Indianapolis
Indiana University School of Medicine, Health &
    Hospitals Corporation of Marion County, Bell-
    Flower Clinic, Sex Problems Clinic.

Iowa
Mount Vernon
The University of Iowa Center for Sexual Growth & Development.

Kansas City
University of Kansas College of Health, Sciences & Hospitals, Center for Sexual Enrichment & Counseling.

Louisiana
New Orleans
River Oaks Psychiatric Hospital, River Oaks Sex & Marital Therapy Program.

New York
Bronx
Albert Einstein College of Medicine, Department of Obstetrics/Gynecology, Division of Human Sexuality.

Brooklyn
Downstate Medical Center, State University of New York, Center for Human Sexuality.

Manhattan
Cornell University College of Medicine, Payne Whitney Clinic, Jewish Board of Family & Children's Service Sex Therapy Clinic, Mount Sinai School of Medicine, Department of Psychiatry, Human Sexuality Program.

New Hyde Park
Long Island Jewish-Hillside Medical Center, Human
Sexuality Center.

New York Medical College, Sex and Marital Therapy
Unit.

New York University Medical Center, Clinic for study
of Sexual Development.

Stony Brook
State University of New York at Stony Brook, School
of Medicine, Sex Therapy Center.

North Carolina
Winston-Salem
Bowman Gray School of Medicine, Marital Health
Clinic.

Ohio
Cincinnati
University of Cincinnati Medical Center, Jewish Hos-
pital, Department of Psychiatry, Human Sexuality
Center.

Cleveland
Case Western Reserve University, Department of Psychiatry, Sexual Dysfunction Clinic.

The Cleveland Clinic, Sexual Dysfunction Unit.

Pennsylvania
Philadelphia
Hahnemann Medical College & Hospital, van Hammett Psychiatric Clinic.

Jefferson Medical College, Department of Psychiatry & Human Behavior, Jefferson Psychiatric Associates Sex Therapy Clinic.

The Medical College of Pennsylvania and Hospital Department of Psychiatry, Psychiatric Clinic & Stress Clinic.

Temple University Medical School, Moss Rehabilitation Hospital.

University of Pennsylvania School of Medicine, Division of Family Study, Marriage Council of Philadelphia.

South Carolina
Charleston
Medical University of South Carolina, Department of Psychiatry & Department of Obstetrics/Gynecology.

Tennessee
Memphis
The University of Tennessee College of Medicine, Department of Psychiatry, Sexual Problems Unit, Sexual Dysfunction Clinic.

Texas
Galveston
University of Texas Medical Branch, Department of
Obstetrics/Gynecology, Sexual Dysfunction Treat-
ment Clinic.

Houston
Baylor College of Medicine, Baylor Psychiatry Clinic.

University of Texas Medical School at Houston, Department of Reproductive Medicine & Biology.

San Antonio
The University of Texas Health Science Center at San
Antonio, Psycho-Social Clinic.

Utah
Salt Lake City
University of Utah Medical Center, Department of
Psychiatry, Sexual Problems Clinic.

Vermont
Essex Junction
University of Vermont College of Medicine, Department of Obstetrics/Gynecology, Sexual Counseling Service.

Virginia
Falls Church
The Johns Hopkins Medical Institutions, Fairfax Hos-
pital, Department of Psychiatry.

Washington
Seattle
University of Washington School of Medicine, University Hospital, Sexual Dysfunction Clinic.

Wisconsin
Milwaukee
Medical College of Wisconsin, Department of Psychiatry.

Sources/Quotations
Bibliography

Abbott, M. A., and McWhirter, D. P. "Resuming Sexual Activity after Myocardial Infarction." *Medical Aspects of Human Sexuality*, 12:18, 1978.

Andersen, B. L. (1983). Primary orgasmic dysfunction: Diagnostic considerations and review of treatment. *Psychological Bulletin*, 93(1), 105–136.

Barnes & Noble, *"Atlas of Human Anatomy."*

Beutler, L. E., Scott, F. B., and Karacan, 1. "Psychological Screening of Impotent Men." *Journal of Urology*, 116:193, 1976.

Blair, L. *Getting Control*, Plune, 1992.

Blaivas, J. G., O'Donnell, T. E., Gottlieb, P., et al. "Comprehensive Laboratory Evaluation of Impotent Men." *Journal of Urology*, 124:201, 1980.

Bommer, J., Ritz, E., del Pozo, E., et al. "Improved Sexual Function in Male Heldialysis Patients on Bromocriptine." *Lancet*, 2:496, 1979.

Bors, E. and Comarr, A. E. "Neurological Disturbances of Sexual Function." *Urological Survey*, 10:191, 1960.

Chopra, D. *Quantum Healing.* Benton, 1989.

Comfort, A. *Sexual Consequences of Disability.* George F. Stickley Company, Philadelphia, 1978.

Cooper, A. J. "A Blind Evaluation of a Penile Ring— A Sex Aid for Impotent Males." *British Journal of Psychiatry.* 124:402, 1974.

Danoff, D. *Superpotency.* 1993.

Davidson, J. M. "Hormones and Sexual Behavior in the Male." *Hospital Practice,* 10:126, 1975.

Deabler, H. L. "Hypnotherapy of Impotence." *American Journal of Clinical Hypnosis,* 19:9, 1976.

De Palma, R. G., Levine, S. B., and Feldman, S. "Preservation of Erectile Function after Aorto-iliac Reconstruction." *Archives of Surgery,* 113:958, 1978.

Divita, E. C. and Olsson, P. A. "The Use of Sex Therapy in a Patient with a Penile Prosthesis." *Journal of Sex and Marital Therapy,* 1:305, 1975.

Dodge, L. J. T., Glasgow, R. E., & O'Neill, H. K. (1982). Bibliotherapy in the treatment of female orgasmic dysfunction. *Journal of Consulting and Clinical Psychology,* 50(3), 442–443.

Ek, A., Bradley, W. E., and Krane, R. J. "Nocturnal Penile Rigidity Measured by the Snap-Gauge Band." *Journal of Urology,* 129:964, 1983.

Fisher, H. *The Sex Contract: The Evolution of Human Behavior.*

Forsberg, L., Gustavii, B., Hojerback, T., et al. "Impotence, Smoking, and B-Blocking Drugs." *Fertility and Sterility*, 31:589, 1979.

Gaylin, W. *Feelings.* 1979.

Gaylin, W. *The Rage Within.* 1984.

Johnson, S. *Sex Is Perfectly Natural.* 1991.

Kaplan, H. S. *The Illustrated Manual of Sex Therapy*, Quadrangle/The New York Times Book Company, New York. 1975.

Kaplan, H. S. *Disorders of Sexual Desire.* New York: Brunner/Mazel, 1979.

Karacan, I. "Advances in the Diagnosis of Erectile Impotence." *Medical Aspects of Human Sexuality*, 12:85, 1978.

Kinsey, A. C., Pomeroy, W. B., Martin, C. E., & Gebbard, P. H. *Sexual Behavior in the Human Female.* Philadelphia: Saunders, 1953.

Kirkpatrick, C., McGovern, K., & Lo Piccolo, J. *Treatment of Sexual Dysfunction.* In G. G. Harris (Ed.), *The Group Treatment of Human Problems.* New York: Grune & Stratton, 1977.

Kramarsky-Binkhorst, S. "Female Partner Perception of Small-Carrion Implants." *Urology,* 12:545, 1978.

Lewis, G. K., & Lo Piccolo, J. (1972). New methods of behavioral treatment of sexual dysfunction. *Journal of Behavior Therapy and Experimental Psychiatry, 3(4),* 265–271.

Little, Blake. *"The Third."*

Lo Piccolo, J. & Lobitz, W. C. (1972). The role of masturbation in treatment of primary orgasmic dysfunction. *Archives of Sexual Behavior,* 2(2), 163–171.

Lyon, H. *Tenderness is Strength.* 1977.

Malloy, Doug. *"Master Piercer."*

Masters, W., Johnson, V., & Kolodory, R. *Heterosexuality,* New York: Harper Collins Publishing, 1994.

Miller, J. S. *The Healing Power of Grief.* 1975.

Mills, L. C. "Drug-Induced Impotence." *American Family Physician,* 12:104, 1975.

Milsten, R. *Male Sexual Function—Myth, Fantasy, Reality,* Avon Books, New York, 1979.

Morokoff, P. Determinants of Female orgasm. In J. Lo Piccolo and L. Lo Piccolo (Eds.), *Handbook of Sex Therapy.* New York: Plenum Press, 1978.

Morokoff, P. & Lo Piccolo, J. (1982). Self-management in the treatment of sexual dysfunction. In P. Karoly & F. H. Kanfer (Eds.), *Self-Management and Behavior Change*. (pp. 489–521), New York: Permanent Press.

Newman, H. F. and Northrup, J. D., "Problems in Male Organic Sexual Physiology," *Urology*, 21:443, 1983.

Offit, A. *The Sexual Self*, Methuen Publications, 1977 & 1983.

Osborne, D. "Psychological Evaluation of Impotent Men." *Mayo Clinic Proceedings*, 51:363, 1976.

Parfrey, Adam. *"Apocalypse Culture."*

Parkes, C. M. Bereavement: *Studies in Grief in Adult Life*. New York: International Universities Press, 1972.

Power, B. *Good Relationships are Good Medicine.*

Rosen, R. C., Shapiro, d.m., and Schwartz, G. E. "Voluntary Control of Penile Tumescence." *Psychosomatic Medicine*. 37:479, 1975.

Schimidt, C. W. and Lucas, M. J. "Impotence." *Primary Care*, 2:275, 1975.

Simon, S. *Forgiveness.*

Singer, J. and Singer I. Types of female orgasm. In J. Lo Piccolo and L. Lo Piccolo (Ed.), *Handbook of Sex Therapy*, New York: Plenum Press, 1978.

Stewart, T. D. and Gerson, S. N. "Penile Prosthesis: Psychological Factors," *Urology*, 7:400, 1976.

Teitleman, J. *The Courage to Grieve.* 1980.

Terman, L. *Psychological Factors in Marital Happiness.* New York: McGraw-Hill, 1938.

Valins, L. *When a Woman's Body Says No to Sex*, New York: Penguin Group, 1992.

Vallely, J. F. "Inflatable Penile Prostheses and Endoscopic Procedures." *Urology*, 13:705, 1979.

Van Thiel, D. H., and Lester, R. "Therapy of Sexual Dysfunction in Alcohol Abusers: A Pandora's Box." *Gastroenterology*, 72:1354, 1977.

Wood, R. Y. and Rose, K. "Penile Implants for Impotence." *American Journal of Nursing*, 78:234, 1978.

Worden, J. W. *Personal Death Awareness.* Englewood Cliffs, N.J.: Prentice-Hall, 1976.

Note: This bibliography makes no pretense at being complete. It is intended as an introductory guide to the interested reader. Because of the complexity of the subject, I have offered a wide choice of titles offering thorough and varied observations.